Heinrich Blitz

Practical Methods for Determining Molecular Weights

Heinrich Blitz

Practical Methods for Determining Molecular Weights

ISBN/EAN: 9783337341633

Printed in Europe, USA, Canada, Australia, Japan

Cover: Foto ©berggeist007 / pixelio.de

More available books at **www.hansebooks.com**

Practical Methods for Determining Molecular Weights

BY

HENRY BILTZ

Privatdocent at the University in Greifswald

TRANSLATED (WITH THE AUTHOR'S SANCTION) BY HARRY C.
JONES, ASSOCIATE IN PHYSICAL CHEMISTRY IN JOHNS
HOPKINS UNIVERSITY, AND STEPHEN H.
KING, M.D., HARVARD UNIVERSITY

EASTON, PENNA.:

The Chemical Publishing Company

1899.

VICTOR MEYER

IN MEMORIAM

TABLE OF CONTENTS

PAGE.

DERIVATION OF MOLECULAR WEIGHT FROM VAPOR-
DENSITY... 1

Theory of vapor-density determination 2

THE GAS-DISPLACEMENT METHOD........................... 6

Description of a simple apparatus for the gas-displacement
method.. ... 8
Carrying out a simple vapor-density determination....... 10
Modifications of the vaporizing vessel 14
Source of heat...... 18
The substance 22
Determination of volume............................... 24
Filling the vaporizing vessel with an indifferent gas...... 27
Measuring the temperature of the experiment........... 32

OTHER METHODS BASED UPON THE GAY LUSSAC PRINCIPLE. 34

Mercury displacement method........... 34
Hofmann's method 35
Method of Dumas....................................... 36

DETERMINATION OF THE DENSITIES OF GASES 43

Determination of vapor-densities under diminished pres-
sure... 44
Procedure of La Coste................................. 45
"   "   Lunge and Neuberg 45
"   "   Dyson, and Bott and Macnair 46
"   "   Schall................................... 47
"   "   Malfatti and Schoop...................... 47
"   "   Habermann 47

CRITICAL EXAMINATION OF RESULTS....................... 48

Smaller deviations 48
Dissociation 49
Difference between the results of the Dumas, and the gas-
displacement method............................ 58

OSMOTIC METHODS....................................... 62

Beckmann's differential thermometer................... 66

DETERMINATION OF THE MOLECULAR WEIGHT BY THE
FREEZING-POINT METHOD 73
The simple freezing-point apparatus of Beckmann........ 79
Carrying out a simple molecular weight determination,
with the Beckmann freezing-point apparatus.......... 83
Mechanical stirring device....... 93
Procedure when hygroscopic solvents are used........... 96

DETERMINATION OF THE MOLECULAR WEIGHT OF SOLIDS ... 100
   The thermostat 102
   The behavior of individual solvents 105
   Table of solvents 106
   Solvents which cannot be used in certain cases 111
   Increase in accuracy in investigating very dilute solutions. 113

CRITICAL EXAMINATION OF RESULTS 117
   Smaller anomalies inherent in the method 117
   Electrolytic dissociation 121
   More complex molecules 125

DETERMINATION OF MOLECULAR WEIGHT BY THE
     BOILING-POINT METHOD 141
   The simple boiling-point apparatus of Beckmann 145
   Carrying out a simple molecular weight determination,
     with the simple boiling-point apparatus of Beckmann. 149
   The boiling-point apparatus of Jones 161
   Carrying out a determination with the Jones apparatus... 164
   Modifications of the boiling-vessel 167
   Modifications of the boiling-jacket 170
   The heating 172
   The introduction of the substance 175
   The use of the different solvents 177
   Some solvents which cannot be used in certain cases 181
   Effect of atmospheric pressure 184

CRITICAL EXAMINATION OF RESULTS 186
   Smaller deviations inherent in the method 186
   Electrolytic dissociation 189
   Complex molecules 191
   Choice of method 196

DETERMINATION OF MOLECULAR WEIGHT FROM THE PRIN-
     CIPLE OF LOWERING OF SOLUBILITY 197

DETERMINATION OF THE MOLECULAR WEIGHT OF
     HOMOGENEOUS SOLIDS OR LIQUIDS 202

DESCRIPTION OF THE METHOD OF TRAUBE 205
   Experimental determination of the molecular volume 206
   Calculation of the molecular volume 206
   Determination of the molecular weight 211

MODIFICATION OF THE TRAUBE PROCEDURE FOR SOLUTIONS. 219
   a. Indifferent solvents 220
   b. Aqueous solutions 221

DETERMINATION OF THE DENSITY OF A LIQUID 225

TABLES 229

AUTHOR'S PREFACE

The methods for determining molecular weights, have been extended and increased within the last few years, by a number of pieces of work. These contributions are scattered throughout the literature.

A brief description of the methods can be found in a number of smaller special works. These books are, however, considered to be somewhat too elementary for the use of a chemical laboratory. They show how, in simple cases, the molecular weight of a substance can be ascertained, and they suffice as a guide for carrying out a determination for practice. The investigations in the chemical laboratory, however, include very often, more complex cases, which can be dealt with only by a thorough study of the original literature. The purpose of this book is to furnish information in such cases, either directly, or by referring to the existing publications.

The forms of apparatus, used by chemists in determining molecular weights, are described here, and as far as appears to be necessary, are shown in drawings. The manner of carrying out the experiments is described with considerable fullness. Details, which facilitate work, practical devices, which teaching and my own experience have furnished, are treated at considerable length. Modifications of apparatus, and manipulation, are treated in special sections. I believe that the experimenter will not have his individual initiative interfered with, by such a detailed description of the methods. As soon as he has overcome the

difficulties first met with, and in doing so this book
will be helpful, he will introduce changes and improvements, in each particular case, according to his own
judgment.   Special stress has been laid upon the thorough study of the results of experiment.   A critical
examination of the observations, which is sufficiently
thorough, is of the greatest importance for the correct
exploitation of the experiments, and certainly, in many
cases, this is not easy.

While I have treated the subject mainly from the
point of view of practice, nevertheless, I could not re-
sist the temptation to add to certain sections, some in-
troductory remarks, which would give a theoretically
clear, and rational explanation of the methods.   These
introductions to the chapters are to be regarded as very
elementary, and should serve rather as incentive to
further study, than as real instruction.

I have included the interesting method of Traube,
which has been only recently discovered, since it will
undoubtedly be of great value to chemists, in many
cases.   The results of this method should be used with
some caution, until it has been thoroughly worked
out and established.

It was my intention to dedicate this book to my hon-
ored teacher Victor Meyer, the brilliant investigator in
the field of molecular science, as an offering on his
fiftieth birthday.   A few weeks before his death, the
proof was submitted to him, and the dedication ac-
cepted by him.   Let the work be now dedicated to
his memory.   PROF. DR. HENRY BILTZ.

Greifswald, August, 1897.

TRANSLATORS' PREFACE

In the preparation of the English edition of this work, we have been guided by the belief, that it would be of service in American and English laboratories, where molecular weight determinations are made. It can be fairly claimed that there is no book in English, and probably not in any other language, which deals with the problem of molecular weight determinations as satisfactorily as this recent publication by Biltz. We believe that the book will not only be useful in the laboratory, but on account of its method of treatment, will also prove to be of service as a book of reference to the literature of molecular weight determinations.

A comparison of the translation with the original, will show that a number of additions, omissions, and changes, have been made. These have either been introduced by the author and forwarded to us, or were made with his approval. We have added an index.

At first we thought of preparing a short chapter on the method of determining the molecular weights of pure liquids, by measuring their surface-tension. We have, however, concluded that the method as worked out by Ramsay and Shields (Ztschr. phys. Chem., **12,** 433), is too refined for general laboratory use, and would, therefore, scarcely find a place in the present volume.

We have tried to avoid a too literal translation. We have endeavored to ascertain the author's meaning, and to express this as clearly as possible, in idiomatic English.   HARRY C. JONES.

                    STEPHEN H. KING.

DERIVATION OF MOLECULAR WEIGHT FROM VAPOR-DENSITY

—

Avogadro[1] advanced the hypothesis in 1811 that the same number of molecules is always contained in equal volumes of different gases at the same pressure and temperature. At the same time he pointed out that a method for the determination of molecular weights can be founded upon this principle. If equal volumes of different gases always contain an equal number of , molecules, then the molecular weights are proportional to the densities of the gases. Since, however, other investigations have shown that the hydrogen molecule consists of two[2] atoms, its molecular weight being therefore two, *the densities of gases referred to hydrogen as unity, when doubled, give directly the molecular weights of the substances in the gaseous state.*

Air is 14.367 times as heavy as hydrogen. The vapor-density referred to hydrogen as unity is obtained from the vapor-density referred to air as unity by

---

[1] A. Avogadro : Ostwald's Klassiker der exakten Wissenschaften, No. 8, pages 3 and 4.

[2] Avogadro furnished the only proof, up to 1868, that the molecule of hydrogen consists of at least two atoms, and it was an assumption that it did not contain more than two. A. Kundt and E. Warburg. (Pogg. Ann., 135, 337 and 527 (1868)), showed that the molecule of mercury is monatomic, whence it follows that the hydrogen molecule is not composed of more than two atoms.

multiplying by 14.367, and the molecular weight by multiplying by 28.73.[1]

On the other hand, the density of a substance referred to air as unity is calculated by dividing the molecular weight, in terms of hydrogen as the unit, by 28.73. If the molecular weight of oxygen is taken as = 16, as is now generally done in accordance with the suggestion of Ostwald, the number 28.95 is to be used instead of 28.73.

It has been the custom for a long time to select air as the unit for vapor-density determinations. Practically this is permissible, but theoretically it is to be rejected, because air is a mechanical mixture and its composition therefore varies, if only slightly. It is desirable to preserve the usual method of calculation, because it has been applied almost without exception to the data thus far obtained.

The vapor-density method for determining molecular weights is indispensable for the solution of a number of questions, especially among inorganic compounds, while it has been replaced in many cases by osmotic methods to be described later, especially in the field of organic chemistry. Vapor-density determinations are indeed of interest at elevated, and at very high temperatures. The most important changes of apparatus, etc., necessary for working at high temperatures, should therefore be described, if only briefly, and the more important literature given.

Theory of Vapor-density Determination.—*By vapor-density is to be understood the density of a substance in*

[1] H. Kopp: Compt. rend., **44**, 1347 (1857).

the gaseous state, compared with air at the same temperature and pressure. The vapor-density is a constant,[1] independent of the temperature, since from the law of Gay Lussac, all gases undergo equal changes in volume for equal changes in temperature.

In order to calculate a vapor-density we must know the weight of the substance (g), and that of the air (G), which, at the same temperature and pressure, occupies the same volume as the substance when vaporized. Then the density d is :

$$d = \frac{g}{G}.$$

The weight of the air G can be ascertained by direct weighing, as Bunsen[2] has done in a series of determinations. It is more convenient to avoid the weighing and to measure the volume instead. In the latter case, however, pressure and temperature must be taken into account. The weight of 1 cc. of air, under normal conditions (0° C., 760 mm.), is 0.001293 gram. The volume V, reduced to normal conditions, is found from the laws of Gay Lussac and Boyle, from the following equation : If v is the volume read at t° C. and

---

[1] The density of a solid or a liquid is dependent upon the temperature, even if water at the same temperature as the substance is taken in every case for comparison, because the coefficient of expansion of water is not the same as that of the substance under investigation. For this reason, water, which is not at the same temperature as the substance, but is at 4° C., is selected for density determinations. The densities thus obtained give the mass of the substance contained in unit volume.

[2] R. Bunsen : Ann. d. Chem. u. Pharm., 141, 273 (1867).

p mm. pressure, and a is the coefficient of expansion of air ($a = 0.00367$),

$$V = \frac{v\,p}{760\,(1 + a\,t)}.$$

From which :

$$G = \frac{0.001293\,v\,p}{760\,(1 + a\,t)}; \text{ and } d = \frac{760\,g\,(1 + a\,t)}{0.001293\,v\,p}.$$

For the logarithmic calculation of a vapor-density determination it is desirable to collect the numerical factors so that their quotient is in the denominator :

$$d \quad \frac{g\,(1 + a\,t)}{0.0000017013\,v\,p}.$$

The table given in the appendix contains $\log \dfrac{1}{1 + a\,t}$ for the values of t, which most commonly occur. To these are added the logarithms of the values in the denominator ($\log 0.0000017013 = 0.23079 - 6$), and the sum is subtracted from $\log g$.

This is the general formula for calculating the vapor-density. It is modified when certain values cannot be determined directly ; $c.\ g.$, if g can be ascertained only through a reduction, as in the method of Dumas.

The formula, modified for such cases, is given in the proper places.

The values necessary for the calculation of a vapor-density are thus: g, v, t, p.

Two different principles are applicable for the experimental determination of the vapor-density of a substance, the one established by Gay-Lussac,[1] the other by Dumas.[2]

[1] L. J. Gay Lussac : Ann. de. Chim., 80, 218 (1811) ; J. B. Biot : Traité de Physique, I, 291.

[2] J. B. Dumas : Ann. Chim. Phys. [2], 33, 341 (1826).

According to the first principle, *a weighed amount of the substance is vaporized, and its volume, in the form of gas, determined*, pressure and temperature being observed. In this case the measurement of the volume is the essential feature. On the contrary, according to the principle of Dumas, *the weight of the substance is determined, which occupies a definite volume* — say that of a small flask — temperature and pressure being also measured. The principle amounts essentially to a determination of weight.

The methods based upon the first principle, *i. e.*, the methods involving gas displacement, are preferable for the use of the chemical laboratory, because a smaller amount of substance suffices, 0.1 to 0.2 gram being enough; whereas methods based upon the principle of Dumas require several grams. This is especially true in the case of new substances, whose molecular weight determination is of chief interest, there being usually only a small amount of substance at disposal. Furthermore, it is far simpler to weigh a given quantity of a substance and then determine its volume as a gas, than the reverse, *i. e.*, to determine the weight of a given amount of vapor.

The first useful method for determining the vapor-density of solids or liquids was that of Gay Lussac, which was discovered in 1811. But it could be employed conveniently only for substances boiling at low temperatures. This explains why it was supplanted by the method of Dumas, published in 1826. The apparatus employed by the latter is simpler, and can be used even at tolerably high temperatures. The method of Dumas was the most important for the determination of vapor-density for almost a half century, indeed, until 1868, when preference was given to the improved form of the Gay Lussac method devised by Hofmann, especially for the investigation of organic substances. Ten years later the gas-displacement method of V. Meyer appeared, and this is preferred to all others, even up to the present day. Determinations by the Hofmann method are carried out now only for practice. The method of Dumas finds occasional application for the solution of certain theoretical questions. The gas-displacement method has replaced both of these for the purpose of determining molecular weights.

THE GAS-DISPLACEMENT METHOD

The gas-displacement method was discovered in 1878, by Victor Meyer,[1] and his pupil C. Meyer. It is the most convenient and elegant method for the determination of the vapor-density of a substance which is solid or liquid at ordinary temperatures. It is always applicable when the substance to be vaporized is not decomposed at the temperature of the experiment, and does not act chemically on the vessel. It has been found to be applicable even at temperatures as high as 1800°. It is less useful than the method of Dumas only for substances which undergo dissociation at the temperature of the experiment, because complications are introduced by the air which is present in the apparatus. The accuracy of the results of the gas-displacement method is somewhat less than

[1] V. and C. Meyer: Ber d. chem. Ges., 11, 1867, 2253 (1878).

that of the other methods, but it is quite sufficient for the demands of the chemical laboratory.

The procedure is as follows: A small weighed portion of the substance (g) is introduced into a vaporizing vessel, which is filled with air and heated above the boiling-point of the substance. As the substance volatilizes, a volume of air equal to the volume of vapor formed, is driven out through the exit tube into the upper colder portion of the apparatus, where it is properly measured (v). The temperature t, at which the volume of the gas is read, and the pressure p, complete the data required for calculating the density.

The temperature which obtains in the vaporizing vessel does not enter into the calculation, and therefore does not need to be known. The volume of air displaced by the vapor, contracts as it leaves the hot vaporizing vessel and passes into the colder portions of the apparatus in which it is to be measured, the amount of the contraction depending upon the change in temperature. Consequently, the calculation reduces the volume to normal conditions only from the temperature and pressure at which the reading is made.

A barometric correction is to be introduced, if, as is usually the case, the volume of air is read over water. The tension of the water-vapor, which depends upon the temperature, produces a corresponding diminution in pressure, so that the volume read over water appears to be greater than it really is. This error is corrected by subtracting from the barometric reading, the tension (f), which the water-vapor exerts at the temperature at which the volume of air is read. The tension can be taken from the table in the appendix. The formula to be employed for the calculation is then:

$$d = \frac{g\,(1 + a\,t)}{0.00000170 13\,v\,(p - f)} .$$

The calculation can be simplified by employing a table which was prepared by G. G. Pond, a pupil of V. Meyer. A number is introduced into the formula for t and p, which, when multiplied by g, and divided by v, gives at once the density.

Description of a Simple Apparatus for the Gas-displacement Method. — The apparatus necessary for a simple density determination by the gas-displacement method, consists of three parts; the vaporizing vessel with top, the device for measuring the gas, and the thermostat. The device for measuring the gas is shown in Fig. 2, and the other two parts in Fig. 1.

The vaporizing vessel (A) is a glass tube about 22 cm. in length and 3 cm. in width, closed below, and terminating above in a glass tube 40 cm. long, and with an internal diameter of 0.5 to 0.6 cm. The drop[1] is fastened upon this as a cover, with a rubber tube and wire ligatures.[2] The upright tube of the drop is just as wide as the neck of the vaporizing vessel. A glass rod is shoved into a tube attached to the side, as shown upon the left of the drawing. A

---

[1] H. Biltz and V. Meyer: Ztschr. phys. Chem., **2**, 189 (1888).

[2] Copper wire, 0.5 to 1 mm. in diameter, which has been heated to redness, is employed for ligatures. This is placed around a rubber tube and fastened by twisting the ends together with flat pliers.

piece of thick-walled elastic rubber tubing is drawn over both, and fastened at both ends with ligatures, so that the rod in the tube can be shoved back somewhat, thereby freeing the upright tube of the drop. The rod is then drawn back to its original position by the elasticity of the rubber tube. The substance in "the chamber," resting upon the glass rod, is, by this movement, made to fall into the vaporizing vessel. The drop is closed above with a small cork, or with a rubber tube into which a glass rod is inserted. A second narrow glass tube, through which the displaced air is conducted to the measuring apparatus, is attached to the side a short distance below this closing device. A glass tube about 2 mm. in diameter, serving as a delivery tube, is bent as shown in Fig. 1, and during an experiment opens under the mouth of the tube for measur-

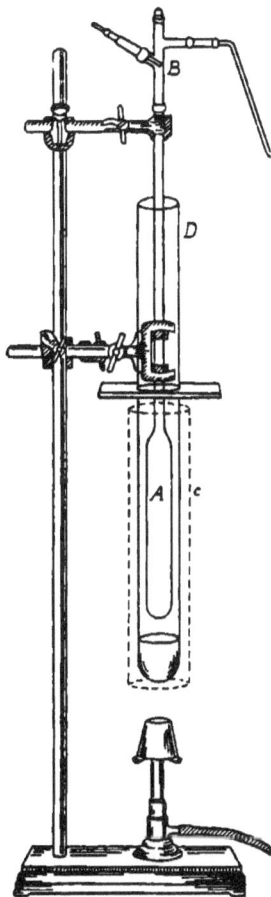

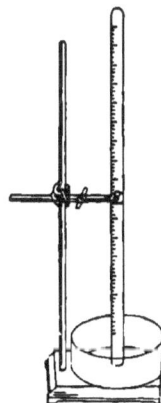

Fig. 2.
Device for measuring
gas. One-tenth natural size.

ing the gas, which stands in a dish as shown in Fig. 2.

The vapors of substances boiling in a vapor-jacket (D), 5 cm. in diameter and 60 cm. in length, and closed like a test-tube, serve for heating the vaporizing cylinder. Through a suitable choice of substances to be boiled, temperatures as high as 600° can be conveniently reached. This form of heating has the advantage that the vaporizing vessel can be heated with certainty to *a temperature which remains constant*, which is of the greatest significance for the method.

The several parts of the apparatus are fastened most conveniently to a large laboratory iron stand, one meter in height, as shown in Fig. 1. This is placed upon a low box, which rests upon the floor close to the table upon which the experiment is being conducted. The arrangement for measuring the gas, and the vessel with water into which this dips, are placed upon the table. The drop and the measuring tube are thus at a convenient height, and the boiling can also be easily observed in the boiling-jacket, and in the interior of the vaporizing vessel.

Carrying Out a Simple Vapor-density Determination.—The first requirement in a vapor-density determination is to clean and dry the apparatus. The vaporizing vessel is cleansed by rinsing with acid, water, alcohol, and ether. It is dried by clamping it horizontally upon a stand, and introducing, quite up to the end of the vaporizing vessel, a small glass tube through which, when attached to a suction-pump, the air is drawn out of the apparatus, while fresh air enters through the neck. The apparatus is uniformly heated by means of the large flame of a Bunsen burner, which is kept in motion beneath it. After the apparatus is dried some asbestos, heated to redness, is placed in the bottom of the vessel, in order to prevent it from being broken by the falling glass flask, during an experiment. The drop is attached to the neck of the vaporizing vessel by means of a piece of very thick-walled rubber tubing (wall is 2 to 3 mm. thick), and fastened by ligatures. The remaining details of the arrangement are given in Fig. 1.

Let toluene be chosen as the substance for the first experiment. About 0.07 gram is introduced into a small weighed tube, about 3 to 4 mm. wide and 15 mm. long, which serves as a weighing tube. The substance is introduced by means of a glass tube which is drawn out and serves as a pipette. During the weighing the little glass vessel is placed in a small stand made of sheet brass. The second weighing is made just before the experiment, or a little tube closed with a glass stopper must be employed (Fig. 3).

Fig. 3. Weighing-glass for substance. One-half natural size.

Forty cc. of aniline, whose boiling-point is approximately 183°, are introduced into the vapor-jacket. The aniline is boiled by means of an ordinary Bunsen burner, which is surrounded by a metal screen to protect it from air currents. The flame is so regulated that the boiling-jacket is filled with the vapor well above the wide portion of the vaporizing vessel, and that condensation begins at several centimeters above this

point. The narrow neck of the apparatus then passes
through the plane in which the vapor comes in
contact with the superincumbent air. This plane
must never lie upon the upper portion of the vaporizing vessel proper, since then the slightest changes
in the height to which the vapor rises,—such as
would be caused by a current of air,—would produce
marked fluctuations in the temperature and corresponding changes in the amount of air contained in
the vaporizing vessel. Such fluctuations which take
place well up in the neck of the vessel, do but little
harm, because of its small diameter. The drop should
be left open during the heating for the same reason,
in order that the water into which the capillary dips
should not be sucked over through it into the
apparatus, in case of an unexpected, sudden cooling of
the latter.

When the temperature is so regulated that the
vapor condenses upon one and the same portion of the
vapor-jacket, without large fluctuations either up or
down, we can proceed with the experiment. The
glass with the weighed substance is slipped into the
chamber of the drop. If a tube with glass stopper is
used, care should be taken that it is closed very
loosely. The use of a stopper should be avoided whenever possible. The drop is now closed. The mouth
of the capillary in the water is observed for several
minutes. If no more bubbles escape from it, it is an
indication that complete temperature equilibrium
has been reached. The determination can be proceeded
with only when this is the case. For this purpose, the

measuring tube, which holds 50 cc. and is graduated to
tenths of a cubic centimeter, filled with water, is placed
over the mouth of the capillary. The rod of the drop
is then drawn back, so that the glass slips into the
apparatus. Vaporization begins. The displaced air
rises bubble by bubble in the measuring tube, and
ceases only when all the liquid toluene in the apparatus has disappeared. Usually a bubble, produced
by the last particles of the substance adhering to the
walls, appears later, and then the experiment is completed. After a few minutes the water begins to rise
in the capillary, due to the fact that the vapors, diffusing high up into the vaporizing vessel, begin to enter
the colder neck and condense there. The drop
should be opened immediately in order to prevent the
water from entering the hot apparatus. The measuring tube, closed with the thumb of the hand in which
it is held, is introduced into a wide glass cylinder of
sufficient height, which is filled almost to the rim
with water at the temperature of the room, so that the
tube is completely immersed. In about five minutes
the gas acquires the temperature of the surrounding
water. The measuring tube is now quickly raised
sufficiently high out of the water, so that the water on
the inside is at the same level as that on the outside,
the measuring tube being held close to the wall of the
cylinder. The meniscus of the water in the tube can
be distinctly seen through the water which has risen
high up between the two glass surfaces, and the
reading can be accurately made. Care must be taken
that in slow reading the measuring tube is not warmed

by the fingers which have raised it from the water. The tube is then either seized with a pair of tongs, or a check reading is made after a few moments. Finally, the temperature of the water and the height of the barometer are read.

Example.—Two experiments gave the following data:

g = 0.0685   p = 774   g = 0.0539   p   774
v = 16.4   t = 5   t = 12.8   t = 5.7

From these data the following vapor-densities and molecular weights are calculated.

	Density.	Molecular weight.
Found	3.258	94.1
	3.263	94.2
Calculated	3.180	92.0

The many modifications of V. Meyer's gas-displacement method, which have been proposed, indicate how widely it can be employed, and the many requirements to which it can be adapted. Useful results can be obtained by means of these, under the most varied conditions, from very low to very high temperatures.

Modifications of the Vaporizing Vessel. — Vaporizing vessels of the dimensions given can be employed up to 600°. It is not necessary to shorten the cylindrical portion, since it can always be heated uniformly. Apparatus made of potassium glass can be employed for still higher temperatures, and can be heated in a gas-furnace or a lead-bath. But porcelain vessels of a somewhat more compact form[1] are more convenient.

---

[1] Total length, 66 cm.; length of neck, 51 cm.; length of cylinder, 14.5 cm.; external diameter of neck, 1.3 cm.; of cylinder, 4.6 cm.; thickness of the walls of the neck, 0.35 cm.; of the cylinder, 0.2 cm. About six weeks are required to prepare the apparatus.

These are made to order at the Royal Porcelain Factory, in Berlin, at eight marks apiece. They are made airtight by glazing both the inner and outer side, and can be used up to 1600°. For still higher temperatures vessels are prepared from "Masse 7,"[1] by the same establishment. It is better to glaze these vessels only on the exterior, because it is difficult to make the inner glazing impervious, and if it is not, a complete drying of the apparatus is impossible. Indeed, a complete drying of an apparatus of "Masse 7," whose interior is unglazed, is difficult, and it is recommended to use the furnace together with the platinum tube devised by myself.[2] There is no material known up to the present which can be used above 1800°, so that this is the highest temperature at which vapor-density determinations can now be made.

Platinum vessels are very useful even up to 1700°, for many investigations, e. g., with the salts of the alkali metals, and can, with some care, be advantageously employed. The price for a weight of say 800 grams is very high, but is not to be considered, because the platinum works are generally ready to take the apparatus back after the experiment, so that only the cost of making the apparatus must be paid. Such platinum apparatus cannot be heated directly in the furnace, because gases from the flame (e.g., hydrogen),

---

[1] The following description of "Masse 7," was received by letter. "Masse 7" is a very difficultly fusible mixture of a silicious to porcelain-like nature. It was discovered by Dr. Hecht in the Royal Porcelain Factory in Berlin.

[2] H. Biltz : Ztschr. phys. Chem., 19, 406 (1896).

diffuse through red-hot platinum. They must therefore be separated from the flame by a tube of porcelain
or " Masse 7."

In many experiments it is convenient to give the
apparatus the form[1] shown in Fig. 4, especially when in the vapor-density determination, the apparatus must be filled with a gas
other than air, as must be done when the
air would act chemically on the vaporized
substance. For example, in the determination of the vapor-density of arsenic, air must
be excluded; in the determination of the
vapor-density of magnesium, nitrogen must
be excluded. The apparatus is filled from
below with an indifferent gas, such as nitrogen or carbon dioxide, through the narrow
entrance tube. It is better to introduce
hydrogen at the top. The stream of gas is
not discontinued until the substance is introduced into the chamber. Every trace of a
foreign gas can, thus, with the greatest certainty, be avoided. The narrow entrance

Fig. 4.
Vessel for de-tube is closed with a rubber tube and a screw
termining va-
por-density pinch-cock. This device allows the tem-
with entrance
tube. One-perature to be conveniently measured ac-
eighth natu-
ral size.cording to the principle of the air thermom-

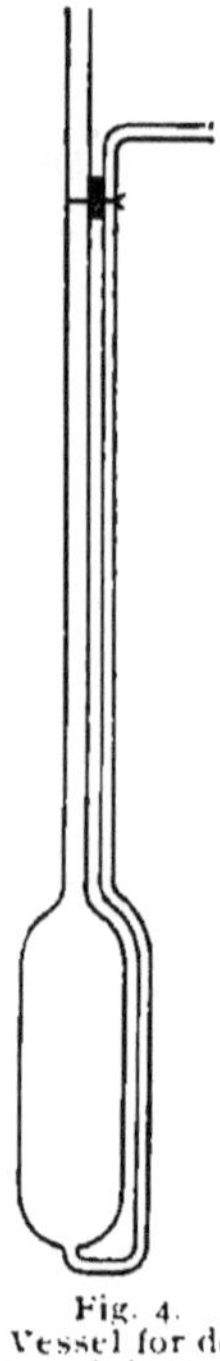

---

[1] Such pieces of apparatus made of platinum, were first used by
J. Mensching and V. Meyer. Glass ones have been recommended,
and are used by me. (Ztschr. phys. Chem., 2, 922 (1888).) They
have been made also of porcelain, but these have not proven to be
satisfactory, because aside from their brittleness, they crack very
easily on warming.

eter, as was done in the communication referred to in the note.

Finally, with this arrangement, it is possible to carry out several vapor-density determinations[1] in succession, with the same apparatus, without removing it from the thermostat. The drop is opened after the first determination, and the apparatus sucked out through the narrow entrance tube by means of a water-pump. The vapors contained in the apparatus are thus removed from it. A condensation of the vapors in the entrance tube is prevented by heating the upper portion of it. When all the vapors have been removed from the apparatus, it can be filled again with dry air or an indifferent gas, and used for a second experiment.

Drops constructed otherwise than the above described, can be employed, *e.g.*, that of Mahlmann,[2] in which the chamber is closed below by a large stopcock with wide bore, or that of L. Meyer,[3] which is easily prepared from some pieces of glass tubing and a cork. That described above appears to have supplanted the others, and is the only one which can be used when, as will be described later, temperature measurements are to be carried out with the same apparatus, both before and after the vapor-density determination, according to the principle of the air thermometer.

---

[1] H. Biltz : Ztschr. phys. Chem., 2, 922 (1888).

[2] H. Mahlmann : Ber. d. chem. Ges., 18, 1624 (1885). Here the drop is sketched.

[3] L. Meyer : Ber. d. chem. Ges., 14, 991 (1880). A sketch is given.

Source of Heat.—The question of the supply of heat must take into account this requirement; the cylindrical vaporizing vessel must be heated perfectly uniformly above and below, to the temperature of the experiment, and maintained at this temperature without change. As already stated, vapor-baths are best adapted to this purpose, since they alone furnish a guarantee that all of the upper portions of the apparatus are heated to the same temperature. These should be used whenever possible. The following substances are recommended as heating liquids.

SUBSTANCES TO BE USED IN THE VAPOR-JACKET.

	Boiling-point.		Boiling-point.
Ether	35°	Methyldiphenylamine	292°
Chloroform	61	a-Naphthylamine	300
Benzene	80	Benzophenone	306
Water	100	Diphenylamine	310
Toluene	110	Phenanthrene	340
Amyl alcohol	130	Triphenylmethane	358
Xylene	140	Mercury	360
Anisol	152	Antimony triiodide	401
Cumene	153	Sulphur	448
Cymene	175	Bismuth tribromide	453
Aniline	183	Phosphorus pentasulphide	518
Nitrobenzene	209	Stannous chloride	606
Ethyl benzoate	213	Stannous bromide	619
Naphthalene	218	Zinc bromide	650
Methyl salicylate	224	Zinc chloride	730
Thymol	230	Cadmium	770
Isoamyl benzoate	262	Zinc	930
Amyl salicylate	270		

When substances are employed which boil above 400°, the vapors cannot generally be driven high up on the neck of the vaporizing vessel. In this case the wide glass tube in which the heating liquid boils, should be surrounded with a sheet-iron tube, which is about 2 ½ cm. wider than the glass tube, as is shown by the dotted line C, Fig. 1. The sheet-iron tube which is held by a clamp fastened to an iron stand, confines the hot gases as they rise, and thus prevents the vapors in the boiling-jacket from condensing too quickly. A screen made of asbestos board is shoved over the boiling-jacket a few centimeters above the upper edge of the sheet-iron tube. The gases rising from the flame are, by this device, deflected away from the upper part of the boiling-jacket. The glass vapor-jackets already described can be used for temperatures up to 600°. Iron tubes must be employed for higher boiling substances; *e. g.*, wide gas tubes forged together at the bottom, or porcelain tubes closed below like test-tubes. These are heated by means of a gas or charcoal furnace.

Since the bottoms of glass vapor-jackets sometimes crack when heated above 200°, vapor-jackets made of glass tubing, open at both ends, can be employed, according to the suggestion of V. Meyer.[1] The lower end is fitted into the grooved collar, which is nearly two cm. deep, of a cast-iron tube, as shown in Fig. 5. In this manner, direct contact of the glass with the flame is avoided. The groove is filled with mercury to close the junction, and for higher temperatures with Woods' metal.

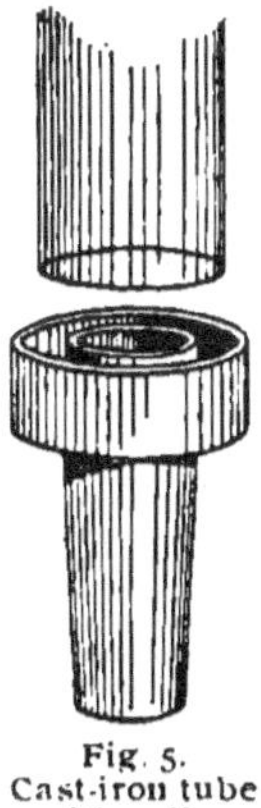

Fig. 5.
Cast-iron tube
with collar.
One-fourth nat-
ural size.

[1] V. Meyer : Ber. d. chem. Ges., **19**, 1861 (1886).

The groove must be filled completely, otherwise the boiling liquid oozes out along the glass wall.

Very low boiling substances require a high vapor-jacket, which is closed at the top by a cork to which a return condenser is attached. In the case of some higher boiling substances, which easily escape or have an unpleasant odor, *e.g.*, naphthalene, naphthylamine, and especially benzoic acid, the vapor-jacket should be closed with a cork to which an upright tube is attached.

Liquid baths are not very useful, since only when constantly stirred do they have an approximately uniform temperature. Baths have been employed containing water, solution of calcium chloride,[1] oil, paraffine, and lead. Baths in which lead is used have the disadvantage that the heavy metal easily presses the vessels in at higher temperatures, and always requires the apparatus to be very firmly clamped.

The furnace devised by L. Meyer,[2] shown in section in Fig. 6 is, on the other hand, very useful. It is heated at *b* by two or more Bunsen burners, whose combustion gases ascending and descending repeatedly in the walls, as in-

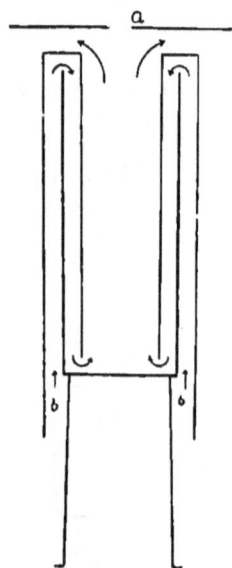

Fig. 6.
Lothar Meyer's furnace.
One-eighth natural size.

[1] Baths of a saturated solution of calcium chloride can be used up to 180°.

[2] L. Meyer : Ber. d. chem. Ges., 13, 991 (1880).

dicated by the arrows, raise the furnace to their own temperature, so that the more elevated portions have approximately the same temperature. If the entire furnace is surrounded by an asbestos screen, and carefully regulated, it maintains the desired temperature quite constant, and this can be measured with a mercury thermometer up to 550°. The drop must be shielded from the hot combustion gases as they ascend, by several screens of foil or asbestos, the lower one being seen at *a*.

Temperatures from 600° to about 1200°, can be most conveniently produced by means of a Perrot[1] gas-furnace, which is very satisfactorily made by the firm of Wiesnegg, in Paris. The same firm also supplies furnaces in which the air used is previously heated,[2] giving temperatures up to 1600°. Instead of the latter an ordinary Perrot furnace can be used, in which a small blast-lamp,[3] consisting of sixteen blast flames issuing from a box, is inserted in the burner circle; or, a blast-lamp[4] obtained by altering the Perrot burner may be used. Temperatures of about 1700° are thus reached. Blast-furnaces in which retort carbon is burned, or better water-gas furnaces,[5] are employed when it is desired to obtain still higher

[1] A good sketch ; Ber. d. chem. Ges., **12**, 1113 (1879), and Wiesnegg's Catalogue, pp. 14 and 15.

[2] Maison Wiesnegg, Paris, 64 rue Gay Lussac. Catalogue, 1894, p. 16.

[3] This procedure was discovered by L. F. Nilson and O. Pettersson : Ztschr. phys. Chem., **4**, 211 (1889). It was moreover used by H. Biltz and V. Meyer : Ztschr. phys. Chem., **4**, 249 (1889).

[4] H. Biltz : Ztschr. phys. Chem., **19**, 388 (1896).

[5] H. Biltz : *Ibid*, **19**, 392 (1896).

temperatures, up to about 1850°. Bellows devised by Root (I use those made at the machine works of Mohr & Federhoff in Mannheim), or by Hoppe, driven by a small gas-engine (Pintsch Brothers, Bockenheim, near Frankfurt, a M.), are used to blow the blast-furnaces. There is no object at present in obtaining a still higher temperature, because we know of no material for vessels which is sufficiently resistant above 1800°.

The Substance.—A small vessel either opened or closed is employed always as described, for the vapor-density determination of liquids. This vessel is made of as thin glass as possible, so that it can be allowed to drop into the apparatus without protecting the bottom of the latter with asbestos. Generally, the asbestos does no harm. Only in some cases can a decomposition of complex molecules be brought about by a foreign body with a rough surface. But in such cases a glass flask with ground stopper is also to be avoided, because the ground portions can also effect dissociation. Tertiary amylacetate,[1] which breaks down into amylene and acetic acid, is an example of this.

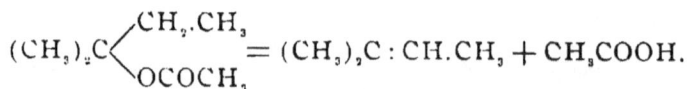

$$(CH_3)_2C\Big\langle{}^{CH_2.CH_3}_{OCOCH_3} = (CH_3)_2C:CH.CH_3 + CH_3COOH.$$

Little vessels of porcelain or platinum are used at higher temperatures, or very thin-walled glass vessels, which are allowed to fuse with the glaze of the porce-

---

[1] N. Menschutkin and D. Konowalow: Ber. d. chem. Ges., **17**, 1361 (1884); Konowalow: Ber. d. chem. Ges., **18**, 2808 (1885).

lain apparatus in the experiment. Vessels of Woods'
metal are used for special purposes. These can be
obtained from the factory of C. Desaga, in Heidelberg,
or they can be made by dipping a smooth cold wire
of 2 mm. diameter into Woods' metal which is just
about to solidify. A vessel is formed around the end
of the wire on solidification, and the walls of this can
be made thinner by paring.

Solid bodies are introduced whenever it is possible,
in compact pieces, without the use of vessels. Such
pieces are cut with a knife either out of a larger piece,
or out of a tablet prepared with the tablet press which
will be described later, and then reduced to the desired
form and required weight by paring.

An easily fusible substance can be readily brought
into the form of sticks, as V. Meyer[1] suggested, by
melting it and drawing it up into a glass tube 2 mm.
wide, and when cold, pressing the stick out with
a wire. If the stick should adhere to the glass in
places, the tube should be slightly warmed, thus
melting the outer layers of the stick adhering to the
glass wall, whereupon the remainder can be easily
thrust out.

Only when solids must be converted into vapor at
a temperature far above their melting-point, should
they be introduced in vessels,[2] in order to prevent the
stick as it slips down, from adhering, while melting,
to the neck of the apparatus. When an apparatus is

---

[1] R. Demuth and V. Meyer: Ber. d. chem. Ges., **23**, 313. Annık.
1890.

[2] V. Meyer: Ztschr. phys. Chem., **6**, 9. Annık. 1890.

employed whose neck has the width already stated, this inconvenience will scarcely be met with.

An important question in the carrying out of vapor-density determinations is, How much substance should be used in an experiment? Even up to 600°, it is recommended to use substance enough to displace 15 to 20 cc. of air. At temperatures above 1200°, not more than 10 cc. of air should be displaced, especially if the substance volatilizes slowly. The weights corresponding to these amounts are obtained by multiplying the vapor-densities expected, by 0.02 (at moderate temperatures) to 0.01 (at high temperatures). When too much substance is used, portions of the vapor rise up into the cold neck, by the diffusion of the gas which takes place very quickly at high temperatures, and condense there. Consequently, a quantity of air which is too small would be displaced, and the vapor-density found, too high.

Determination of Volume. — A measuring tube with a capacity of 50 cc. and graduated to tenths of a cubic centimeter, is employed in the ordinary vapor-density determination, as described. Since we are not certain that these tubes are calibrated for gas-measurements (they are more frequently calibrated with liquid), the following procedure is advisable[1] in the case of very exact measurements. Introduce some air-bubbles into the tube before the displaced air is received, and ascertain their volume in the way indicated. After the deter-

[1] V. Meyer : Ber. d. chem. Ges., **25**, 631 (1892).

mination is completed the total volume is read in the usual manner. The difference is the volume of the gas which has been driven out, provided the temperature is the same at both readings, otherwise a correction is necessary. This procedure has the advantage that in both cases a meniscus reading is made, and we are therefore independent of a calibration not adapted to the purpose.

If a large glass cylinder is not at disposal the measurement of the volume can be made in the simplest manner, thus: The measuring tube with its volume of gas, as shown in Fig. 2, and the liquid into which it dips are placed where the tube will not be exposed to great changes in temperature. After a quarter of an hour the temperature in the neighborhood of the cylinder and the volume are read, and the height of the column of water from the level of the outer liquid to the meniscus in the tube are measured with a ruler. The number of millimeters is divided by 13.6, and the quotient subtracted from the barometric height.

Sometimes it is convenient to employ a gas-burette,[1] Fig. 7, e. g., when it is feared that the water will run back; further, when substances which are very easily vaporized, are to be investigated at high temperatures. In this case vaporization generally proceeds so rapidly that some bubbles emerge at the side of the tube and are lost. The gas-burette consists of a U-tube with one limb calibrated, and surrounded by a water-

[1] The gas-burette was first proposed for vapor-density determination by Fr. Meier and J. M. Crafts: Ber. d. chem. Ges., **13**, 856 (1880).

jacket.   This limb is attached above to a thick-walled
capillary tube.   The U-tube is attached at its lower
part to a reservoir, *b*, for the liquid which is to fill it,
by means of a rubber tube.   The gas-burette is con-

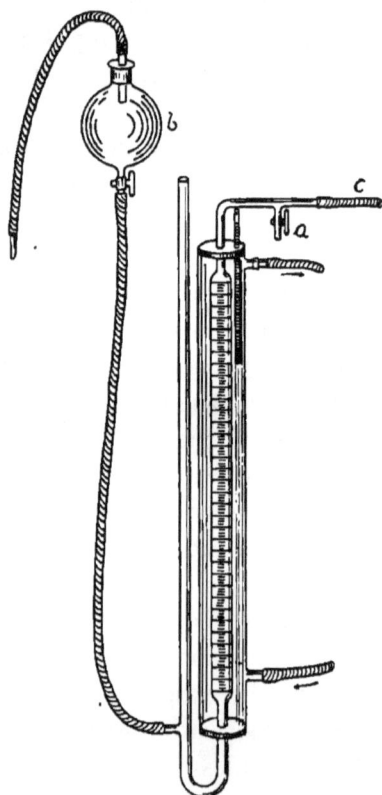

Fig. 7. Gas-burette. One-fifteenth natural size.

nected at C, with a vaporizing vessel, through a glass
or lead tube, whose internal diameter is about 2 mm.
The stop-cock, *a*, remains open during the heating.
The surface of the liquid in the measuring tube is

raised to the uppermost division, with the aid of the reservoir, *b*. After closing *a*, the determination of the density is proceeded with in the usual manner. *b* is lowered until the liquid in the measuring tube stands at the same level as the liquid in the outer limb, before the reading is made.[1] Nilson and Pettersson[2] have employed a gas-burette filled with mercury, which necessitates the use of a manometer tube filled with a lighter liquid.

Instead of using a gas-burette in the case of rapid vaporization, it is also practicable to receive the gas first in a glass tube $2\frac{1}{2}$ cm. wide, narrowed below to about 1 cm., and which is drawn out above and provided with a rubber tube closed with a pinch-cock.

After the gas has been collected in the tube the latter is completely immersed under water, and by carefully opening the stop-cock the gas is allowed to escape from the wide tube into a narrower measuring tube held over it.

Filling the Vaporizing Vessel with an Indifferent Gas.—Those substances on which the oxygen of the air would act chemically, such as arsenic, sulphur, benzaldehyde, or those which react with nitrogen, such as magnesium, require the vaporizing vessel to be filled with an indifferent gas. Carbon dioxide can be employed, but is not to be recommended on account

---

[1] H. Biltz: Ztschr. phys. Chem., **4**, 251 (1889); **19**, 411 (1896). A complete apparatus for the determination of vapor-density and for the measurement of temperature, at very high temperatures, is sketched here.

[2] L. F. Nilson and O. Pettersson : J. prakt. Chem., N. F., **33**, 1 (1886).

of its solubility in water. Hydrogen is better adapted
to the purpose. It is best to select nitrogen. This is
prepared as follows: Air which has been freed from
carbon dioxide by sodium hydroxide, is passed through
very concentrated ammonia, and, saturated with
ammonia, is passed over red-hot copper turnings. The
oxygen of the air combines with the hydrogen of the
ammonia forming water, and pure nitrogen leaves the
tube. The gas is further purified by a repetition of
this process, and, finally, to test whether there is any
oxygen present, the gas, free from ammonia, is passed
over red-hot copper into the vaporizing vessel. If no
oxygen is present the copper will remain completely
untarnished. The technical details of this method,
which are to be recommended especially when larger
quantities of nitrogen are used in a series of investigations, are published in the *Zeitschrift für physikalische
Chemie.*[1]

If only a little nitrogen is required it is prepared
according to Böttger,[2] by gently warming equal parts
of sodium nitrite, ammonium chloride, and potassium
bichromate, with three parts of water. Care must be
taken, since a too violent evolution of gas can easily
take place. A large generating flask is always to be
used, and the flame removed from the flask as soon as
the reaction begins.

The vaporizing vessel can be filled most conveniently and with the greatest certainty, with the appa-

---

[1] H. Biltz : Ztschr. phys. Chem., **19**, 409 (1896).

[2] Böttger : Jahresbericht des physik. Vereins zu Frankfurt
a M., 1876-'77, 24.

ratus provided with entrance tube (Fig. 4). If such is not available the gas is introduced through a narrow glass tube, which passes through the neck of the apparatus for determining vapor-density, and reaches clear to the bottom. The stream of gas is not to be interrupted during the removal of the tube, after the apparatus has been filled. After the tube is removed it is held over the drop while the substance is introduced into the chamber, to prevent air from diffusing in, and then the apparatus is immediately closed.[1]

A further advantage, which can also be useful in an ordinary vapor-density determination, is secured by filling the apparatus with hydrogen. It is possible in this way to carry out vapor-density determinations of substances near, and even below their boiling-point, as V. Meyer and R. Demuth[2] have shown. The rapidly diffusing hydrogen helps to convert the substance into vapor, even below its boiling-point, because the partial pressure of the substance is small in the resulting gas-mixture.

It was thus possible to determine the vapor-density of xylene which boils about 140°, at 100°, and that of ethyl ether (boiling-point 35°) at the temperature of the room.

For this procedure, it is understood that the hydrogen should have the freest possible access to the substance to be vaporized. For this purpose vaporizing vessels with flat bottoms are employed, so that the

---

[1] A still more certain method of doing this was described by H. Biltz : Ztschr. phys. Chem., 19, 412 (1896).

[2] R. Demuth and V. Meyer : Ber. d. chem. Ges., 23, 311 (1890).

liquid can spread out over these in a thin layer. And
further, when the chemical nature of the substance
permits, vessels of Woods' metal should be used.
These melt at from 60°–80°, and on fusing empty
their contents. These vessels must be as thin walled
as possible, so that they will not break through the
bottom of the vaporizing vessel. This should not be
protected in this case with asbestos, glass wool, etc.,
because it would absorb the substance and make its
vaporization difficult. At most, some small spirals of
platinum wire can be used. When necessary, short,
wide, glass vessels can be used instead of those made
of Woods' metal. In this case the pouring out of the
contents must be facilitated by gently tapping the
apparatus.

Other gases, such as nitrogen and air, can also be
present in the apparatus in this procedure, as Krause
and V. Meyer[1] have shown. The vaporization then
requires more time, because these gases have smaller
diffusion velocity. Only a small quantity of substance
can be taken for an experiment, for reasons already
given. The determination of the vapor-density of
sulphur was made successfully at its boiling-point, in
an apparatus filled with nitrogen.

When chlorides are to be investigated the vaporizing vessel is filled with chlorine, and a gas-burette
filled with concentrated sulphuric acid is used for
measuring the volume. The displaced chlorine would
be too soluble in water. If a gas-burette is not avail-

[1] A. Krause and V. Meyer: Ztschr. phys. Chem., 6, 5 (1890).

able, a pipette[1] of about 150 cc. capacity, filled with
air, and having a narrow entrance tube, is introduced
between the drop and the capillary leading to the
measuring tube. The pipette is kept at a constant
temperature, and this is accomplished most conveniently, by surrounding it with a glass jacket through
which steam is conducted. The chlorine issuing from
the apparatus displaces an equal volume of air from
the pipette, and this is then measured in the usual
manner. A glass three-way piece, of small bore, is
introduced between the pipette and the drop. This
connects the apparatus first with the hood or an absorption vessel, but during the experiment itself the
way to the pipette is left open.

In connection with this method, the attempt has been
made to prepare the chloride, whose vapor-density was
to be determined, right in the vaporizing vessel itself.[2]
A diminution or an increase in volume takes place,
depending upon the composition of the chloride
formed. From the magnitude of this change the
vapor-density of the chloride[3] can be calculated. A
gas-burette must be used for this purpose, because in
consequence of the heat which is set free in the

---

[1] W. Grünewald and V. Meyer: Ber. d. chem. Ges., 21, 698
(1888).

[2] H. Biltz: Ber. d. chem. Ges., 21, 2766 (1888).

[3] This method was employed for determining the vapor-
densities of indium chloride and ferric chloride, and by filling the
apparatus with oxygen, of sulphurous acid. This method has
recently been employed by A. Vandenburg: Ztschr. anorg. Chem.,
10, 58 (1895), for the determination of the vapor-density of
molybdenum dioxychloride.

reaction, there is at first a larger increase in volume than corresponds to the reaction.

Measuring the Temperature of the Experiment.— It is not necessary to know the temperature at which the experiment is carried out in order to calculate the density as determined by the gas-displacement method. In many cases, however, it is desired to know the temperature, especially when dissociating vapors are investigated, therefore in the case of such substances which give different values for the vapor-density at different temperatures. In order to ascertain in this case the dependence of the density upon the temperature of the experiment, the latter must be measured.

The problem is simple when the vapors of boiling liquids whose boiling-point is known, can be employed, or liquid or air-baths whose temperature can be measured with the mercury thermometer.

The measurement of high temperatures is difficult, especially when an electrical thermometer is not available.[1] The method of the air-thermometer is the safest in this case, and is the only one which can be employed in measuring temperatures above 1700°. Here the vessel itself in which the density determination is made, serves as the air-thermometer. The measurement can be carried out according to either of

---

[1] Thermoelectric pyrometers, which can be employed even above 1600°, have recently been put upon the market by Heraus, in Hanau, and Keiser and Schmidt, in Berlin. These pyrometers are based upon the magnificent pieces of work of L. Holborn and W. Wien, in the physical technical "Reichsanstalt." Ann. d. Phys. und Chem., N. F., **47,** 107 (1892); **56,** 360 (1895).

three principles: First, the amount of air contained in
the apparatus, whose volume at 0° had been previously ascertained, is determined at the temperature to
be measured, by expelling it by means of another gas,
such as carbon dioxide or hydrochloric acid which is
easily absorbed by the liquid beneath, and then measuring it.[1] Second, according to the method of Regnault the amount of gas which leaves the apparatus
during the heating is collected, and the amount of air
remaining in the apparatus is calculated from this.[2]
Third, a method which is essentially adapted to physical investigations consists in measuring the increase
in pressure in the apparatus, the gas contained in the
apparatus being prevented from escaping. In order
that too great pressure should not be produced at high
temperatures, which would injure the apparatus, the
heating is begun with the apparatus partially ex-
hausted, a pressure of about an atmosphere being
reached at the temperature of the experiment.[3]

The details of this method can be found in the
papers referred to.

[1] J. M. Crafts and Fr. Meyer: Compt. rend., **90**, 606 (1880) ;
H. Goldschmidt and V. Meyer: Ber. d. chem. Ges., **15**, 141 (1882) ;
J. Mensching and V. Meyer: Ztschr. phys. Chem., **1**, 152 (1887) ;
H. Biltz and V. Meyer: Ztschr. phys. Chem., **2**, 188 (1888).

[2] H. V. Regnault: Mém. d. l'academie royale des sciences de
France, **21**, 163 (1847); H. Biltz and V. Meyer: Ztschr. phys.
Chem., **4**, 252 (1889); H. Biltz : Ztschr. phys. Chem., **19**, 385 (1896).
The same method with certain modifications was employed by L. F.
Nilson and O. Pettersson : J. prakt. Chem., N. F., **33**, 1 (1886).

[3] L. Holborn and W. Wien : Ann. d. Phys. und Chem., N.
F., **47**, 125 (1892).

OTHER METHODS BASED UPON THE GAY LUSSAC PRINCIPLE

There are several methods for determining vapor-density based upon the Gay Lussac principle, which are included under the general head of "mercury-displacement methods." They are mentioned here only incidentally, because they rarely find application at present.

A tube, bent in the form of a U, with one limb closed, can be employed, as A. W. von Hofmann and V. Meyer have proposed. The substance floats in the tube on mercury, which completely fills both limbs of the tube. The substance is converted into vapor by heating the entire apparatus above the boiling-point of the substance in question, the temperature being accurately determined, and the volume is calculated from the weight of the mercury which flows out of the open limb. The difference between the heights of the mercury in the two sides of the U-tube, is added to the barometric reading. Since, in the experiment, the entire mass of mercury is heated to the tempera-ture of the experiment, the tension of the mercury vapor, which corresponds to the temperature of the experiment, is subtracted from the pressure. (Mercury displacement method of Hofmann and V. Meyer.[1])

---

[1] A. W. v. Hofmann : Ann. Chem. und Pharm. Suppl., 1, 9 (1861). V. Meyer: Ber. d. chem. Ges., 10, 2068 (1877). The for-mula to be used in the calculation is given on page 2071.

Woods'[1] metal is used instead of mercury at higher temperatures. A. W. v. Hofmann[2] determined the volume, later, in a somewhat different manner, by a subsequent filling with mercury, as given in the exact description of his method very similar to the gas-displacement method.

The Hofmann[3] modification of the Gay Lussac method, which consists in vaporizing the substance in the vacuum of a mercury barometer, was very much used in his time. The volume of the vapor is read directly on the calibrated tube. The height of the mercury column is measured with a cathetometer, and after introducing a correction both for temperature and tension of the mercury vapor, it is subtracted from the barometric reading. Since, in this method, a more or less diminished pressure exists in the apparatus, according to the amount of substance employed and the size of the apparatus used, the determination of the density can be carried out below the boiling-point of the substance in question; e. g., the vapor-density of cumarine, which boils at about 270°, can be determined at 183°.

This method has the disadvantage that only the upper portion of the mercury column in the tube is heated, while the shorter lower portion is not. The

---

[1] V. Meyer: Ber. d. chem. Ges., **9**, 1216 (1876).

[2] A. W. v. Hofmann: *Ibid.*, **11**, 1684 (1878).

[3] A. W. v. Hofmann: *Ibid.*, **1**, 198 (1868); **9**, 1304 (1876).

pressure which the mercury column exercises, reduced to 0°, cannot, therefore, be calculated from this alone. This is made possible by the change proposed by Wichelhaus,[1] which makes it possible to heat the entire mercury column uniformly to the temperature of the experiment, or by Hofmann's[2] modification, in which the mercury tube is closed below, after the substance is vaporized, and the mercury column is measured only after cooling.

It is possible to use the Hofmann method up to 300°, through a modification proposed by Brühl.[3] At higher temperatures, the tension of the mercury is, to a very marked extent, dependent upon the temperature. Brühl has succeeded in eliminating the tension from the formula by introducing a measurement of the tension into the experiment, and including in the calculation the difference between the height of the mercury column in the determination of tension and in the experiment proper. The use of the method of Brühl is of value, especially for temperatures above 240°.

METHOD OF DUMAS

The method of Dumas[4] for determining vapor-density is based upon the Dumas principle, men-

---

[1] H. Wichelhaus : Ber. d. chem. Ges., **3**, 166 (1870).

[2] A. W. v. Hofmann : *Ibid*, **9**, 1306 (1876).

[3] Brühl : *Ibid.*, **12**, 198 (1879).

[4] J. B. Dumas: Ann. Chim. Phys., [2] **33**, 341 (1826)

tioned in the introduction to this chapter It consists in determining the amount of substance in the form of vapor which a flask, whose volume is to be subsequently ascertained, will hold. This method is, in general, less convenient to carry out than those already described, and therefore seldom finds application at present in the chemical laboratory. It is, however, indispensable for the solution of certain

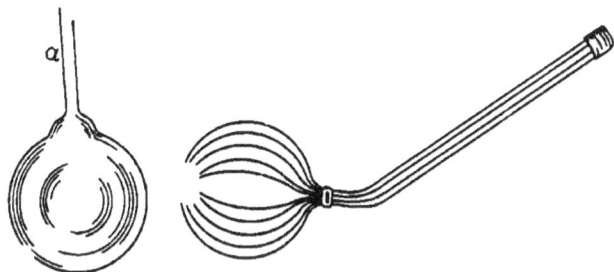

Fig. 8.
Dumas bulb with holder.   One-fourth natural size.

problems, as will be shown later, and therefore should be more fully described.

The apparatus for the ordinary determinations of low-boiling substances is unusually simple. A balloon flask holding from 100 to 250 cc. is employed. Its neck, close down to the bulb, is drawn out to a tube about 5 mm. in diameter, Fig. 8. The flask is carefully cleansed, dried just like the vaporizing-vessel in the gas displacement method, and then weighed

empty.[1] Four or five grams of the substance are
introduced, and the neck of the flask, 4 or 5 cm. from
the bulb, is drawn out to a capillary. The piece of
glass tube thus separated from the flask is weighed,
and its weight subtracted from that of the empty
flask. In determining the vapor-densities of liquids
the tube of the dried flask in which the density
determination is to be made, is drawn out before the
first weighing, then the flask is weighed, and finally
the amount of substance required is drawn in by
slightly warming and cooling the flask.

The flask, in the determination proper, is seized by
the wire holder shown in Fig. 8, and brought into a
thermostat, which consists of a vapor- or water-bath, or
one in which calcium chloride is used. The temperature of the bath must be at least 10° above the boiling-
point of the substance to be investigated. The substance, as it vaporizes, drives the air out of the flask,
and the larger portion of the substance escapes with it
through the capillary. When the substance is completely volatilized, which is easily recognized in that

---

[1] In order that the flask may be weighed with sufficient exactness, a flask of about the same size, and only a little lighter, is
hung on the other side of the balance, and the difference in weight
determined. The same flask is used as a tare in the subsequent
weighings of the density bulb. By this device, we are made independent of the disturbance which depends upon the height of the
barometer, temperature of the weighing room, humidity of the air,
etc. V. Regnault: Compt. rend., 20, 975 (1845); J. prakt. Chem.,
35, 206 (1845).

the stream of vapor ceases and a small drop of the
condensed substance continually closes the mouth of
the capillary, the capillary is fused shut by means of
a small, portable blast-lamp,[1] which is blown with the
mouth through a rubber tube. The flask is now
removed from the bath in which it was heated,
cleansed, and weighed closed. Then the closed point
is scratched with a glass knife (a file would remove
too much glass), and broken off under water which
has been boiled and thus freed from air, or under
mercury. The flask becomes completely filled, with
the exception of a small bubble of air, which is generally not driven out with the vapor of the substance.
The flask is now weighed, then the space occupied by
the air-bubble is filled and a second weighing is made.
The weighings should be made accurately to centigrams with water, and to decigrams with mercury.

The data necessary for the calculation are:

m, weight of the empty open flask.

m′, weight of the closed flask filled with vapor.

M, weight of the flask completely filled with water
or mercury.

M′, weight of the flask filled up to the bubble with
water or mercury.

t, temperature of the bath when the capillary is
fused together.

b, height of the barometer at this moment.

t′, temperature of the flask filled with vapor, when
weighed.

---

[1] C. Desaga : Heidelberg. Katalog, 1888, Nr. 349.

b′, height of the barometer during the weighing of the flask filled with vapor.

λ, density of the air[1] when the flask filled with vapor is weighed. ·

3 β, cubical expansion coefficient of the glass = 0.000025.

Q, the density of the medium[2] employed to fill the flask and determine its volume.

Then we have:

$$d = \frac{(m' - m)\ \dfrac{Q}{\lambda} + M' - m'}{(M - m)\ \dfrac{1\,(b + at')}{b'(1 + at)}\ [1 + 3\beta(t - t')] - (M - M')}$$

This formula is taken from the "Leitfaden der praktischen Physik," by Kohlrausch, page 69, where the formula is also derived.

It is better in many cases to determine the amount of substance contained in the flask not in the manner given, but by means of chemical analysis, thus avoiding considerable sources of error. Iodine and arsenious acid can be easily and accurately determined by titration. The contents of the flask are rinsed out with a suitable solvent, and the amount of substance ascertained by analysis. The calculation must be somewhat modified in this method of procedure. The

---

[1] The value of λ is to be taken from the table given in the appendix.

[2] The value of Q is to be taken from the table given in the appendix. It is generally sufficient to take Q 1 for water, Q 13.56 for mercury.

formula necessary for this purpose is published in the Zeitschrift für physikalische Chemie.[1]

Ostwald[2] recommends that the substance be condensed in the point, by cooling this and warming the closed flask. This portion is then fused off, and the weight of the substance ascertained, by weighing this portion first with the substance and then alone. The substance can be weighed more accurately in this way than was possible in the original procedure of Dumas. It was difficult to make the weighings of the flask with sufficient accuracy by the latter method.

If, in the Dumas method, vapors of boiling substances are employed as means of heating, it is advisable to produce them in a somewhat shortened glass-jacket, having the same form as the jacket described in the gas-displacement method, and to employ cylindrical vessels[3] for the density determination, instead of round flasks. Notwithstanding their elongated form they are heated uniformly by the surrounding vapors. The vessel in which the density determination is made, is held by a wire or glass support, and so suspended in the boiling-jacket that the fine drawn-out point protrudes somewhat beyond the jacket. This portion is warmed during the experiment by means of a heated metal ring, held by a handle, and moved to and fro over it, to prevent the substance from con-

---

[1] H. Biltz : Ztschr. phys. Chem., **19,** 421 (1896).

[2] W. Ostwald : Hand- und Hilfsbuch zur Ausführung physiko-chemischer Messungen, p. 125 (1893).

[3] O. Pettersson and G. Eckstrand : Ber. d. chem. Ges., **13,** 1191 (1880).

densing in the point and closing it. The density is calculated by the usual Dumas formula.

Nilson and Pettersson[1] have shown, in their magnificent work on the determination of the density of aluminum chloride, how this method can be used for substances which are hygroscopic and easily decomposed.

The method of Dumas is not adapted to higher temperatures. The experimental errors, due to the difficulty of an exact measurement of temperature, accumulate to such an extent that the result is generally worthless. The vessel must be made of some material other than sodium glass for temperatures above 600°.[2] Potassium glass can be used up to about 750°. Deville and Troost[3] have carried out determinations at still higher temperatures with porcelain vessels. Their results have subsequently been shown to be faulty,[4] at least in part. And this is not to be wondered at, since the flask in which the density is determined, contains at 900° to 1000° only such a small amount of the substance in the form of vapor, that it is extremely difficult to determine the exact weight of the substance contained in the flask during the experiment. Density determinations made by the Dumas method, at all events above 700°, are not relable. (Determination of the density of sulphur by Bineau.[5])

[1] L. F. Nilson and O. Pettersson : Ztschr. phys. Chem., 4, 217 (1889).

[2] H. Biltz : Ibid., 2, 934 (1888).

[3] H. St. Claire Deville and L. Troost : Ann. Chim. Phys., [3], 58, 257 (1860).

[4] H. Biltz : Ztschr. phys. Chem., 19, 415 (1896).

[5] A. Bineau : Mémoires de la classe des sciences de l'Academie de Lyon, 10, 69. Compare Ann. Chim. Phys., [3], 59, 456 (1860).

DETERMINATION OF THE DENSITIES OF GASES

The Dumas method, whose employment in the investigation of liquids and solids has just been discussed, has been used in various modifications for determining the density of gases. The amount of the gas contained in the apparatus is determined accurately, either as in the unaltered procedure by direct weighing of the vessel, or a vessel is employed which is provided with an entrance tube, as shown in Fig. 4. It is, however, desirable that both of the tubes leading from the cylinder should have thick walls and be of narrow bore. The vessel is filled at the temperature of the experiment with the gas to be investigated. This is then displaced by means of an indifferent gas, and collected in a suitable absorption medium. In this way the weight of the substance can be ascertained by a simple weighing, or by analysis. The weight of an equal volume of air is ascertained by driving it out with hydrochloric acid gas,[1] and measuring it directly, after the hydrochloric acid gas has been absorbed by water. The weight of the air is calculated from the volume. The density in question is obtained by dividing the weight of the gas by the weight of the air.

This method was first used by Marchand.[2] He absorbed oxygen by means of red-hot copper, and determined the increase in the weight of the tube filled with the copper. Carbon dioxide and sul-

---

[1] Carbon dioxide was used at very high temperatures, and absorbed by potassium hydroxide.

[2] R. F. Marchand : J. prakt. Chem., 44, 38 (1848).

phurous acid were absorbed by potassium hydroxide. This method has been very much used at high temperatures by V. Meyer.[1] Chlorine was absorbed by a solution of potassium iodide and determined by titration. Hydrogen was displaced by hydrochloric acid gas and measured as gas. This same method has been employed also at very low temperatures, to determine the density of the hydrogen acids of the halogens.[2]

DETERMINATION OF DENSITIES UNDER DIMINISHED PRESSURE.

The attempt has been made to carry out density determinations under diminished pressure, because at high temperatures vapor-density determinations are not easy to carry out, and furthermore, because, in addition to other reasons, many substances undergo decomposition near their boiling-point. The diminution of pressure makes it possible to work at a lower temperature.

This was the reason why the Hofmann method was in vogue for a long time, although it is not easy to carry out, and could be employed only up to about 200°. Yet, by means of it, the vapor-densities of substances boiling about 100° higher could be determined.

It was attempted to so modify the methods of Dumas and V. Meyer that they could be used under diminished pressure. It must, however, be observed

[1] V. Meyer: Ber. d. chem. Ges., **13**, 399, 2019 (1880). Pyrochemische Untersuchungen, Braunschweig, 1885.

[2] H. Biltz: Ztschr. phys. Chem., **10**, 354 (1892).

that these methods, which for the most part have been
ingeniously thought out, have thus far contributed
almost nothing to an extension of our knowledge.
For this reason only the typical changes have been
considered, and these but briefly.

La Coste's Mode of Procedure.[1]—In this mode of
procedure the V. Meyer apparatus is used. The new
feature is this: The whole apparatus, the vaporizing
vessel as well as the measuring tube with the liquid
into which it dips, are, so to speak, in a space under
diminished pressure, which, however, remains constant
during the experiment. The volume of the gas which
enters the tube at the slight pressure of the experiment,
is brought to the pressure of the atmosphere after the
experiment, read, and introduced into the calculation
in the usual manner, taking into account the pressure
of the atmosphere and the temperature of reading.
The procedure of Th. W. Richards[2] is very similar,
differing somewhat only in the apparatus used.

The Method of Lunge and Neuberg[3] offers the
same advantage, that of working at a pressure already
known. The vaporizing vessel in the V. Meyer
apparatus is connected with an arrangement similar
to the Lunge gas volumeter, which is represented in
general by a gas-burette filled with mercury, on which
a peculiar device makes it possible to read the
gas volume reduced directly to normal conditions
(0°, 760 mm.). This "gas volumeter" serves at first

[1] W. La Coste: Ber. d. chem. Ges., **18**, 2122 (1885).
[2] Th. W. Richards: Chem. News, **59**, 39 (1889).
[3] G. Lunge and O. Neuberg: Ber. d. chem. Ges., **24**, 729 (1891).

as a mercury air-pump for producing the desired pressure in the apparatus, while at the same time the vaporizing vessel is heated to the temperature of the experiment. When this is accomplished the substance is introduced into the vaporizing vessel, and care is taken that the same pressure which exists before the experiment is re-established in the apparatus, by changing the position of the mercury level. The connection between the gas-volumeter and the vaporizing vessel is now broken, and the volume of the air in the volumeter (which is equal to the amount of vapor formed) is reduced to normal conditions and read.

This procedure has been somewhat simplified by Traube,[1] and improved to this extent, that the volume of the displaced air is read at the slight pressure of the experiment, whereby it can be more accurately determined. The pressure which obtains in the apparatus is then introduced into the calculation. The calculation of a vapor-density by the V. Meyer procedure is so simple, that a reading of the volume reduced to normal conditions, as is affected by the Lunge gas-volumeter, is not an essential improvement.

Procedure of Dyson, and Bott and Macnair.—The method of Dyson,[2] that of Bott and Macnair,[3] and also that of Schall, described later, are to be distinguished from the preceding methods in that instead of measuring the increase in volume at constant pressure, the

---

[1] J. Traube : Physikalisch-chemische Methoden, p. 34 (1893).

[2] G. Dyson : Chem. News, **54**, 88 (1887).

[3] W. Bott and D. S. Macnair : Ber. d. chem. Ges., **20**, 916 (1887).

change in pressure at constant volume is measured. It thus resembles the Hofmann method. The apparatus is extraordinarily simple. The vaporizing vessel of a V. Meyer apparatus, with the drop, is connected with a manometer. The desired pressure is established in the apparatus by means of an air-pump. The pressure existing in the apparatus is read before and after the introduction of the substance, and the density is calculated, making use of the known volume of the apparatus.

The Procedure of C. Schall[1] avoids a special determination of volume; the difference in pressure, caused by the vaporization of the substance, being compared with the increase in pressure produced by a small known volume of air which was left in the apparatus in a previous experiment. Either this small volume of air is read directly on a measuring tube, or the volume of carbon dioxide is employed, which is evolved by sulphuric acid from a weighed amount of sodium carbonate, and which can therefore be easily calculated.

H. Malfatti and P. Schoop[2] have described a device which makes it possible to employ the mercury-displacement method of Hofmann and V. Meyer under diminished pressure.

Habermann[3] has shown that the Dumas procedure

----

[1] C. Schall : Ber. d. chem. Ges., **22**, 140 (1889) ; **23**, 919, 1701 (1890).

[2] H. Malfatti and P. Schoop : Ztschr. phys. Chem., **1**, 163 (1887).

[3] J. Habermann : Ann. d. Chem., **187**, 341 (1887). It has been found possible to use this method to determine the vapor-density of indigo.   E. V. Sommaruga : Ann. d. Chem., **195**, 307 (1879).

can be employed under diminished pressure, without essentially changing either the apparatus or mode of procedure. The neck of the Dumas flask passes air-tight into a tube with several bulbs, in which the vapor of the substance, issuing from the flask, can condense. The other end of this tube leads to the manometer, and to a device which allows a constant pressure to be produced. The determination is carried out in the usual manner, and the pressure read on the manometer is introduced into the calculation. A pump run by water, and a regulator constructed by L. Meyer on the principle of the Mariotte flask, and modified by Städel and Hahn,[1] is employed to produce constant pressure.

All methods carried out under diminished pressure have the advantage that a smaller amount of substance suffices. They require larger vaporizing vessels, having a capacity as great as 500 cc., and the work with them must be done very exactly, since otherwise the experimental errors are quite considerable.

THE RESULTS

1. **Smaller Deviations.**— All vapor-density methods give inaccurate values just above the boiling-point of the substance.[2] These deviations are explained thus: Vaporization near the boiling-point is not complete

[1] W. Städel and E. Hahn : Ann. d. Chem., **195**, 218 (1879).

[2] The dilution of the vapor of a substance with another gas, is equivalent to a diminution of pressure or a lowering of the boiling-point. This explains how V. Meyer and A. Krause (Ztschr. phys. Chem., **6**, 5 (1890)) were able to make valuable vapor-density determinations at the boiling-point of the substance.

and therefore a smaller volume of gas is formed than corresponds to the amount of substance. The existence of these anomalies has been proved by thorough investigations, which, however, have not succeeded in ascertaining accurately the factors at work.[1] The rule in actual practice is to carry out a vapor-density determination at least 25° to 30° above the boiling-point of the substance, and to be quite sure of the result the determination should be repeated at a still higher temperature. *The vapor-density determination is free from objection when the two results agree to within the limits of error of the method.*

2. **Dissociation.**—A still greater anomaly was encountered in a number of cases in carrying out vapor-density determinations, which, for a long time, shook confidence in the reliability of the calculation of molecular weights from vapor-density. In some cases, instead of the value of the density expected, only half of this was found; e.g., Bineau[2] found 0.89 for the vapor-density of ammonium chloride, while the formula, NH_4Cl, required the value 1.85. In other cases the density was shown to be dependent upon the temperature, within certain limits; a decrease in density corresponding to a rise in temperature. Thus, Cahours[3] found the following values for acetic acid, whose calculated vapor-density is 2.07, using the Dumas method.

---

[1] Compare W. Ostwald : Lehrb. allg. Chem., I, 161 (1891).

[2] A. Bineau : Ann. Chim. Phys., [2] 68, 441 (1838).

[3] A. Cahours : Ann. Chem. und Pharm., 56, 176 (1845).

Vapor-density of Acetic Acid between 125° and 338°.

Temperature	125°	140°	160°	190°	219°	250°	300°	338°
Vapor-density	3.20	2.90	2.48	2.30	2.17	2.08	2.08	2.08

If the temperatures and the densities corresponding to these are plotted in a curve, this dependence of the vapor-density upon the temperature is quite apparent.

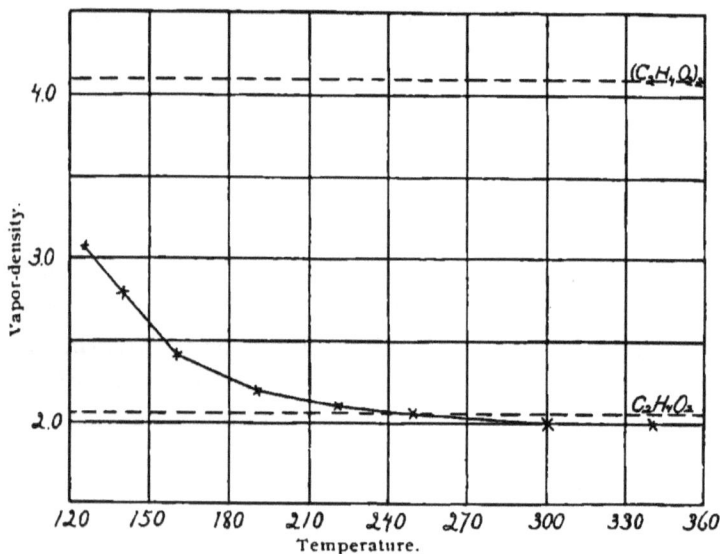

Fig. 9.
Vapor-density of acetic acid as determined by the Dumas method.

V. Meyer[1] on the one hand, and Crafts and F. Meyer[2] on the other, obtained similar results with iodine. The density 8.8 to 8.7 was found from 200° to 600°, from which the formula $I_2 = 8.77$ is derived. A decrease in the density was manifested above 600°. With increasing temperature the density further

---

[1] V. Meyer: Ber. d. chem. Ges., **13**, 394, 1010 (1880).
[2] J. M. Crafts and Fr. Meyer: Compt. rend., **92**, 39 (1881).

decreased, and from about 1400° the vapor-density was again constant, being, however, only about one-half of the above value, 4.5. The gas-displacement method was used in this investigation.

The anomalies first mentioned are due to the same cause. This is clearly seen from the last example. The molecule of iodine even up to 600°, evidently consists of two atoms. These molecules cannot exist at a higher temperature. They decompose into part molecules I_r. At first only a small number of the molecules, I_2, are decomposed. This number increases as the temperature rises, and from 1400° up, all the molecules of I_2 are decomposed to I_r.

Deville[1] has introduced the term "dissociation" for this decomposition of complex molecules. He observed this phenomenon in other investigations. It was a peculiar fate that Deville was indeed the most violent opponent of the application of the dissociation theory for the explanation of anomalous vapor-densities. It was Cannizzaro,[2] Kopp,[3] and Kekulé,[4] who recognized almost simultaneously, that abnormal vapor-densities can be explained by means of this theory.

The experimental proof of a dissociation of the vaporized substance can be very clearly furnished for at least some substances. Among these is phosphorus pentachloride, PCl_5, which gives values for the vapor-density considerably lower than 7.20 which corre-

[1] H. St. Claire Deville : Compt. rend., 45, 857 (1857).

[2] S. Cannizzaro: Nuovo Cimento, 6, 428 (1857).

[3] H. Kopp : Am. Chem. und Pharm., 105, 390 (1858).

[4] A. Kekulé : Ibid.. 106, 143 ; Annik., 1858.

sponds to the formula PCl_5. A. Neumann,[1] using the
Dumas method, obtained the value 5.08 at 182°, at
higher temperatures lower values, and from 290° up,
the constant value 3.7 was found. That a dissociation
had, in fact, taken place according to the equation,

$$PCl_5 = PCl_3 + 2Cl,$$

giving the double number of molecules, which would
explain the half-value for the vapor-density, was shown
by the color of the vapor. This, according to the
investigations of Deville,[2] had the yellow color of
chlorine; being an especially striking example of the
phenomenon of dissociation.

Pebal,[3] in a beautiful diffusion experiment in which
a diaphragm was used, succeeded in separating the products of dissociation of ammonium chloride,—hydrochloric acid and ammonia. The important diffusion
experiments of Wanklyn and Robinson[4] show, perhaps,
in a more striking way, the dissociation of certain
vapors. They vaporized the substance to be tested, in
a flask. The more rapidly diffusing products of dissociation were carried away in larger amounts than the
more slowly diffusing, by the air-current which was
allowed to pass over the neck of the flask. When the
experiment was interrupted after a time, analysis
showed that, as a matter of fact, a larger quantity of

----

[1] A. Neumann: Ann. Chem. und Pharm., Suppl., **5**, 349 (1867).
[2] H. St. Claire Deville: Compt. rend., **62**, 1157 (1866); Ann.
Chem. und Pharm. **140**, 168 (1866).
[3] L. v. Pebal : Ann. Chem. und Pharm., **123**, 199 (1862).
[4] J. A. Wanklyn and J. Robinson : Compt. rend., **52**, 547 (1861);
J. prakt. Chem., **88**, 490 (1863).

the more slowly diffusing products of dissociation was contained in the flask. In the case of phosphorus pentachloride, more chlorine than phosphorus trichloride was carried away, so that when the experiment was interrupted a mixture remained behind which was richer in phosphorus and poorer in chlorine.

The tendency to dissociate was especially manifested by those compounds which are easily formed by bringing together the products of their dissociation; $e.g.$, the ammonium salts and organic ammonium compounds, chloral hydrate, chloral alcoholate, sulphuric acid, nitric acid, sulphuryl chloride, and some alcohols which, through a splitting off of water, form unsaturated hydrocarbons.

The values of the densities found for all of these substances near their boiling-point, lie between the values calculated from the formula and those calculated from the sum of the dissociation-products. At higher temperatures values are obtained which indicate complete dissociation, and these remain constant at still higher temperatures.

The vapor-density values which are higher than would be expected from the formula of the substance, can be explained in a strictly analogous manner; $e.g.$, the vapor-densities found for acetic and formic acids near their boiling-points. It has long been assumed that complex molecules, $\left(\begin{matrix} H \\ COOH \end{matrix} \right)_2$ and $\left(\begin{matrix} CH_3 \\ COOH \end{matrix} \right)_2$ exist here. Only very recently have the different methods for determining molecular weights shown

that this assumption is correct, as will be pointed out
in discussing the results of the freezing-point and
boiling-point methods.   This anomaly disappears in
the case of the fatty acids of higher molecular weight,
butyric acid, etc., both in the vapor-density determination and in the investigation in solution.   The
same relations obtain with nitrogen dioxide as with
the simple fatty acids.   It gives higher values for the
vapor-density at lower temperatures.   The existence
of double molecules has also been shown here with
the freezing-point method.[1]

The more complex molecules have, in some cases, a
greater stability, so that concordant values are obtained
for their vapor-densities within a wider range of temperature.   The density of the vapor of iodine between
250° and 600° points constantly to the formula I_2, that
of the anhydride of arsenious acid,[2] from 520° to 770°,
to the formula As_4O_6.

A constant density would undoubtedly be obtained
again for iodine above 1600°, which would correspond
to the second molecular weight I_1.   But the simple
experiment bearing upon this point has, however, not
yet been made.

It is interesting to note that the dissociation does
not extend to all of the molecules, at one temperature ;
e.g., the vapor-density of arsenious acid at 770°
corresponding to the formula As_4O_6, might be found
at a little higher temperature (say 800°), to correspond
to the formula As_2O_3.   Experiment has shown, indeed,

[1] W. Ramsay : Ztschr. phys. Chem., 3, 66 (1889).
[2] H. Biltz : *Ibid.*, 19, 417 (1896).

that a rise in temperature of over 1000° is necessary to complete the dissociation, a vapor-density corresponding to the formula As_2O_3 not being obtained until a temperature of 1800° was reached.

The kinetic theory of gases furnishes an explanation of this phenomenon, which at first appears so remarkable. According to this theory the temperature of a gas is the expression of the mean velocity of its molecules. Some molecules, indeed, have this velocity, others move with greater, while others again with smaller velocity. The velocity of the different molecules is continually changing and has a different value at each moment. Only the mean velocity of all the molecules remains constant — equality of temperature being assumed. A definite amount of internal energy is necessary to decompose a molecule. Only a limited number of all the molecules present possess this amount at low temperatures. A larger proportion of the molecules obtains the amount of internal energy sufficient for their dissociation with rise in temperature, and the consequent increase in the total velocity of all the molecules. Therefore, the amount of dissociation increases. A constantly increasing number of molecules obtains the energy necessary for their decomposition as the temperature rises, until finally all of the molecules have reached this condition and the dissociation is complete.

If we designate the temperature at which an isolated molecule would be decomposed by dissociation, as the dissociation temperature, it follows that the dissociation temperature is not the temperature at which all the molecules of a gas are really decomposed, but the temperature at which exactly half of the molecules is decomposed. Only in this case do the dissociation temperature and the mean gas temperature coincide. This point would be in the neighborhood of 1400° for arsenious acid.

The "degree of dissociation" is an expression for the extent to which the dissociation has progressed — how large a fraction of the molecules originally present in the undecomposed condition, is decomposed by dissociation. Let d be the density which corresponds to the undecomposed substance, d' the density found at a definite temperature (smaller than d), n the number of part molecules into which each original molecule decomposes (= 2 for arsenious acid), and γ the degree of dissociation, then

$$\gamma = \frac{d - d'}{(n - 1)\, d'}.$$

It follows from what has been said that we must be very careful in judging a vapor-density determination. Several vapor-density determinations must always be carried out at different temperatures, except in simple cases as with organic substances, where a single determination of the density suffices to establish the molecular weight. Only when these different determinations give a constant value, are we justified in drawing a conclusion in reference to the composition of the molecule corresponding to it. It can also occur, especially with inorganic substances, that in addition to the one kind of molecule, a second kind may exist at much higher or lower temperature. The

latter are to be regarded as decomposition or condensation products of the first. Examples are iodine and
arsenious acid.

In a number of cases, as acetic acid, nitrogen dioxide, sulphur, it is possible to prove the presence of
only one kind of molecules,—the smaller,—by means
of the vapor-density method, because the more complex molecules begin to decompose already at the
boiling-point, and dissociation values are obtained just
above this temperature. The method of vapor-density
determination does not suffice in this case to ascertain
the molecular weight of the more complex molecules.[1]

Other methods must be utilized here to furnish
unquestionable proof of their existence. This has,
indeed, been accomplished for the substances named
above, and many others, by the aid of the freezing-
point and boiling-point methods. These methods
have shown that there are molecules of acetic[2] acid
having the composition $(CH_3COOH)_2$, of nitrogen
dioxide[3] having the composition N_2O_4, and of sulphur[4] S_8.

The osmotic methods just named are, in turn, often
not able to demonstrate the existence of smaller molecules, as with iodine, sulphur, and arsenic trioxide.
These methods establish in these cases only the
molecular weights corresponding to the formulas
I_2, S_8, and As_4O_6.

[1] H. Biltz: Ztschr. phys. Chem., 3, 228 (1889).
[2] E. Beckmann: Ibid., 2, 729 (1888).
[3] W. Ramsay: Ibid., 3, 66 (1889); E. Beckmann: Ibid., 5, 80 (1890).
[4] J. Hertz: Ibid., 6, 358 (1890); H. Biltz: Ibid., 19, 425 (1896).

The neglect of this consideration easily leads to error, as is shown by the assumption of the existence of the molecules S_6, which prevailed for many years. The question has not yet been solved whether double molecules always exist in many important cases, such as the chlorides of iron and aluminum, together with the simple molecules $FeCl_3$ and $AlCl_3$. Their existence is, however, indicated by the higher vapor-densities found at lower temperatures. This question has not yet been answered by other methods in a manner which is free from objection. The more recent experiments of Werner are not adapted to this purpose, since the boiling-point method was used.

3. **Difference between the Results of the Dumas and the Gas-displacement Method.**—As far as the determination of the molecular weight of an un-dissociated substance is concerned, the method of Dumas and the gas-displacement method give results of equal value, so that when very accurate measurements are not required the gas-displacement method of V. Meyer is to be preferred, since it is the more convenient. It is different for dissociating substances. Here the two methods give different values; those found by the method of Dumas being uniformly higher than by the method of V. Meyer.

The reason for this lies in the fact that in the gas-displacement method the foreign gas mixed with the vapor dilutes it, producing thereby a corresponding increase in the dissociation. The determinations of the vapor-density of sulphur,[1] carried out by the two

—

[1] H. Biltz : Ztschr. phys. Chem., **2**, 920 (1888).

methods at the same temperature, show this difference
very clearly.

Temperature.	Dumas Method.	Gas-displacement Method.
518°	7.04	5.4 mean
606°	4.73	3.6 "

The results[1] for aluminum chloride obtained by
Nilson and Pettersson confirm the foregoing.

The dissociating influence of a foreign gas mixed
with the vapor of the substance, manifests itself in
still another way. A more or less intimate mixture
of the vaporized substance with the foreign gas takes
place in the V. Meyer method, according to the
amount of substance which is to be converted into
vapor. A more intimate mixture is obtained when
the amount of substance used is small, than when a
larger quantity is employed.

Consequently, in the former case the dissociation
proceeds further than in the latter, and a smaller
value for the vapor-density is found. The fluctuations
caused thereby in the density values of dissociating
substances are, in general, not great, as e. g. is shown
by the experiments of Nilson and Pettersson[2] with
aluminum chloride. They found for aluminum
chloride, at the boiling-point of sulphur :

Gram substance	0.1102	0.0963	0.0859
Density	7.79	7.5	7.4

This dependence of the vapor-density upon the
amount of substance employed, is very unusually

---

[1] L. F. Nilson and O. Pettersson: Ztschr. phys. Chem., 4, 206,
224 (1889).

[2] L. F. Nilson and O. Pettersson : Ibid., 4, 214 (1889).

great for sulphur[1] within its stages of dissociation, since the breaking down of the molecule S_8 into molecules of S_2, produces a marked change in the density.

Gram substance	0.1067	0.0888	0.0675	0.0555	0.0450
Density	7.1	6.4	5.7	4.9	4.5

The form[2] of the vaporizing vessel in the gas-displacement method, has also an influence in the investigation of dissociating substances. The wider the vessel the better the mixing of the vaporized substance with the gas contained in it; the narrower the vessel, the less perfect the mixing. As a matter of fact, methylene bromide at 100°, and sulphur at 518°, behave similarly in this respect.

All these observations show that the Dumas method is better adapted to the investigation of dissociation phenomena than the gas-displacement method. The disturbing factors above mentioned disappear and much simpler relations obtain. But the experimental difficulties presented by the Dumas method above 700°, are so great that the gas-displacement method must be resorted to. This latter method also gives uniform values at these high temperatures, and still better at the very highest temperatures employed, because the movement of the molecules, as such, is so rapid at these high temperatures that an intimate mixing[3] of the gases is, in every case, quickly brought

---

[1] H. Biltz: Ztschr. phys. Chem., 2, 926, 944 (1888); Ber. d. chem. Ges., 21, 2013 (1888).

[2] H. Biltz: Ber. d. chem. Ges., 21, 2772 (1888).

[3] Consequently the dissociation will always increase, for an equal rise in temperature, more rapidly at higher temperatures than at lower. A dissociation curve must therefore fall more rapidly at very high than at moderate temperatures, as is shown by the dissociation curve of arsenious acid (Ztschr. phys.Chem., 19, 422 (1896)).

about. Values are indeed found, which differ from
those which would be obtained by the Dumas method,
but these remain constant in repeated experiments, so
that the dissociation can be followed step by step by
the gas-displacement method.

OSMOTIC METHODS

There is the tendency on the part of a dissolved substance to distribute itself uniformly throughout the entire volume of the solution, just as a gas tends to fill uniformly the space placed at its disposal. The osmotic experiments of Graham[1] and others, in which a layer of the pure solvent was placed over the concentrated solution, have shown this; the dissolved molecules migrating from the solution into the pure solvent, and the reverse, the molecules of the pure solvent wandering into the solution. The final condition which is reached, however, only after a long time, is one in which a sufficient number of molecules have wandered from the solution into the solvent to make the concentration in both equal. The measure of this tendency of the dissolved molecules to fill the space uniformly is the *osmotic pressure*.[2]

Osmotic pressure has been measured directly with suitable devices. These measurements have shown that, apart from temperature and concentration, osmotic pressure is dependent only upon the number of dissolved molecules, but not upon their size. Therefore, equimolecular solutions, *i. e.*, solutions

---

[1] Th. Graham: Ann. Chem. und Pharm., **77**, 56, 129 (1851); **80**, 197 (1851); **121**, 1 (1862).

[2] J. H. van't Hoff: Ztschr. phys. Chem., **1**, 481 (1887).

which contain an equal number of molecules of any substance in equal volumes, have the same osmotic pressure at the same temperature.

Equimolecular solutions are defined differently, as Raoult and others have done, in investigations in which measurements are to be made. Equimolecular solutions are those which contain an equal number of molecules in an equal mass of the solvent. By using this definition the calculation of the results of experiment is considerably simplified, so that at present it is almost exclusively used. Abegg[1] has tried to furnish a rational basis for it.

A method for determining molecular weights is based upon this. Two solutions of different substances in the same solvent are brought to the same osmotic pressure.[2] It follows from what has been said, that in the two solutions an equal number of molecules of the two substances are present in the same amount of the solvents. If we designate the molecular weights by m_1 and m_2, and the number of grams of the substance dissolved in 100 grams of the solvent by r_1 and r_2:

$$\frac{m_1}{m_2} = \frac{r_1}{r_2}.$$

If now the molecular weight of one of the dissolved substances, m_2, is known, the unknown molecular

---

[1] R. Abegg: Ztschr. phys. Chem., 15, 248 (1894).

[2] Compare the work of H. de Vries: Ztschr. phys. Chem., 2, 415, 430 (1888); and of K. Schreber: Mitteilungen aus dem naturwissenschaftl. Verein f. Neupommern und Rügen, 26, 161 (1895).

weight of the other is calculated from the equation:

$$m_1 = \frac{m_2 \, r_1}{r_2}.$$

But since it is difficult to carry out a direct measurement of the osmotic pressure, an indirect method is
employed. Instead of ascertaining the pressure which
a dissolved substance exerts in its effort to fill the
larger volume occupied by the solvent, the work
(osmotic work) is determined in a solution of definite
concentration, which is necessary to separate the
solvent and the dissolved substance. Nernst[1] has
shown that there are several possible ways of doing
this, and how these may be used as methods for
measuring osmotic pressure.

Only three of these find practical application in the
laboratory as methods for determining molecular
weights: (1) The freezing-point method; (2) the
boiling-point method; and (3) the method of Nernst,
based upon the principle of lowering of solubility.
The first is based upon the fact that when a solution
freezes only the solid solvent separates, while the
liquid solution becomes more concentrated. But
since a part of the solvent is thus removed from the
solution and the dissolved molecules are confined to a
smaller volume, the osmotic pressure of the dissolved
substance must be overcome by external work. More
heat must therefore be removed from a solution before
it freezes, than from the pure solvent. Inasmuch as
the latent heat of fusion remains nearly constant, the

[1] W. Nernst: Theoretische Chemie, p. 121 (1893).

larger removal of heat is expressed in a depression of the freezing-point.

A separation of a part of the solvent from the solution also takes place when a solution is boiled, and upon this the second method rests. A special expenditure of work is necessary also here, to confine the dissolved substance to a smaller quantity of the solvent. More heat must be added than is necessary to boil the pure solvent, and this larger addition of heat is expressed in a rise of the boiling-point, since the heat of vaporization is only slightly changed.

The third method attempts to remove the solvent from the solution in a manner different from those already described, by adding another solvent which dissolves the primary solvent to a certain extent, but does not dissolve the substance in solution in the primary solvent. A part of the solvent is removed from the primary solution by means of this second solvent, and the more of it the smaller the osmotic pressure of the dissolved substance in the primary solvent.

The real theoretical foundation of these methods, which could be only indicated here, is discussed more fully in the text-books of physical chemistry.[1] In the following presentation of the several methods, each is built up for itself on an experimental foundation, and is empirically developed.

---

[1] W. Ostwald: Lehrb. d. allg. Chemie (Leipzig, 1891). W. Ostwald: Grundriss d. allg. Chemie (Leipzig, 1889). W. Nernst: Theoretische Chemie vom Standpunkte der Avogadroschen Regel und der Thermodynamik (Stuttgart, 1893).

Beckmann's Differential Thermometer

Since we are not dealing with absolute measurements of temperature in determining molecular weights by the boiling-point and freezing-point methods, but only determined how much higher a solution boils than the pure solvent, or how much lower it freezes, a differential thermometer with an arbitrary scale is employed for this purpose. The numbers on the scale do not indicate the real temperature, but the difference between the two positions of the mercury column, read on the scale, corresponds to the difference in temperatures in degrees Celsius.

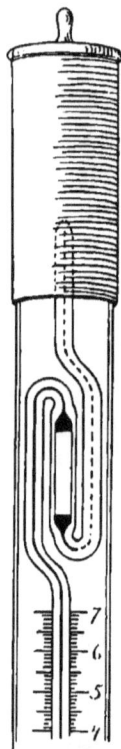

Thermometers adapted to this purpose have been constructed by Beckmann.[1] They are furnished in complete form by F. O. R. Götze, in Leipsic. There are two forms upon the market: those having a larger mercury bulb, about 5 cm. long and 0.9 cm. in diameter, and those having a smaller mercury bulb, about 2.3 cm. long and 1.1 cm. in diameter. The length of a degree is, in the first case, about 4.5 cm., in the latter about 3.5 cm. The scale includes from 5 to 6 degrees. Each degree is divided into hundredths, and each hundredth is sufficiently large that the thousandths can be estimated accurately by means of a lens. Care must be taken in reading, that the eye, the mercury meniscus,

Fig. 10.
Reservoir of the Beckmann differential thermometer. Natural size.

---

[1] E. Beckmann: Ztschr. phys. Chem., **2**, 643 (1888).

and the line to be read, are all on the same plane,
that parallax may be avoided.

Thermometers with smaller bulbs are generally
employed at present, because they can be used for
investigations with both the freezing-point and boiling-
point methods, while thermometers with long bulbs
can be used only up to 80°. However, when it is
desired to procure a thermometer, attention should be
paid to what is stated later (note page 69).

A small reservoir is joined to the capillary above, as
shown in Fig. 10, in order that the same thermometer
can be used at different temperatures. If observations
are to be made at a lower temperature, in the neighborhood of 0°, the bulb of the thermometer must
contain much mercury, and only a small residue is
left in the reservoir. But if the same thermometer is
to be used at about 100°, a considerable quantity of
mercury must be transferred from the bulb of the
thermometer to the reservoir. In determining
molecular weights by the boiling-point method, in
which a rise in boiling-point is to be observed, the
adjustment must be so made that the top of the
mercury column comes to rest on the lower part of
the scale, when the thermometer is immersed in the
vapor of the pure solvent; while in the freezing-
point method, where a depression of the freezing-
point is to be observed, the mercury should come to
rest near the upper end of the scale, when the thermometer is immersed in the freezing solvent. It is a
matter of indifference, as already stated, at what particular number of the scale the meniscus comes to

rest.   It is only necessary that the mercury should come to rest upon the scale, and that a sufficient space is available for the further readings.

The adjustment of the thermometer requires some practice.   There are essentially three modes of procedure.   It is either a question of driving mercury over from the thermometer proper into the reservoir, or the reverse, from the reservoir into the thermometer. The latter can be accomplished in either of two ways.

In order to transfer mercury from the thermometer

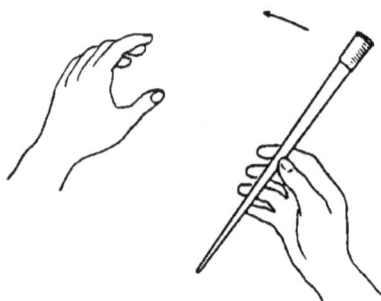

Fig. 11.   Shaking off a drop of mercury.

proper to the reservoir, the bulb of the thermometer is carefully warmed until the mercury flows through the capillary, enters the upper part of the reservoir, and collects here as a hanging drop.   When the necessary amount of mercury has entered the reservoir, the thermometer is seized as shown in Fig. 11, with the tips of the fingers of the right hand, and by bending the wrist is gently tapped against the index finger of the left hand.   The hanging drop of mercury is thus made to fall of, and it then unites

with the mass of mercury already in the reservoir.[1]
If more mercury than is already present should be
introduced into the thermometer proper, the following
manipulation is employed. The thermometer is
seized below, about the lower end of the scale, and its
upper part is thrust downward through the air, as
shown in Fig. 12. In this way the drop of mercury

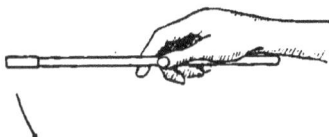

Fig. 12. Throwing a drop of mercury to the top of the reservoir.

resting on the bottom of the reservoir is thrown up to
the top, and unites here with the mass of mercury
which has been driven up through the capillary by
warming the bulb. On cooling, the mercury enters
the capillary. When a sufficient mass has entered,
the drop which still remains suspended, is thrust to
the bottom of the reservoir by the movement first
described.

------

[1] The mercury in the top of the reservoir can be shaken off
most easily with thermometers made recently by Götze, especially
for the boiling-point method. In these the reservoir is wider above
than below, and the capillary is set on, not conical but flat
(E. Beckmann : Ztschr. phys. Chem., 21, 252 (1896), so that a
drop suspended from above clings less firmly than in a cylindrical
reservoir. *These thermometers are not to be recommended for
freezing-point determinations*, because the drop of mercury
belonging to the thermometer proper, and which often appears in
the top of the reservoir during the carrying out of a series of observations, falls off too easily. This should be borne in mind in ordering, and an instrument should be required which is adapted also
to freezing-point determinations.

When it is desired to bring only a small quantity of the reserve mercury into the capillary—as is often the case,—the procedure can conveniently be somewhat modified. Some of the mercury is easily brought from the bottom of the reservoir to the very top, and into the empty capillary above, by repeatedly carrying out the first movement. By repeating the above process this small amount can be driven further into the capillary, and the amount can be increased, the mercury entering only the capillary itself and the extreme upper end of the reservoir, while larger drops are, of course, thrown back into the reservoir by the thrusts given the thermometer. When a sufficient quantity of mercury has collected in the upper end of the capillary, it is united with the mass of mercury in the bulb of the thermometer, by warming the latter until it has flown through the capillary and connected with the former.

In the actual adjustment of a thermometer, these manipulations are to be made use of as follows:

If the thermometer is to be adjusted for a freezing-point determination, the entire amount of mercury is connected (manipulation 2), and the thermometer is then brought into a wide test-tube, which contains 10 to 15 cc. of the solvent in question. The liquid is cooled to freezing. After the mercury has acquired the temperature of the liquid, the thermometer is quickly withdrawn from the liquid, and the excess of mercury is shaken off by manipulation 1. Still the small mass of mercury which is contained in the uppermost uncalibrated part of the capillary must be removed. The bulb of the thermometer is carefully warmed for this purpose, until a small drop of mercury enters the reservoir above, and this is also shaken off (manipulation 1). The thermometer is then reintroduced into the freezing solvent, and it is observed whether the adjustment is such that the end of the mercury column comes to rest upon the uppermost part of the scale. If this has not been accomplished the operation is repeated, or, if too much mercury has been removed, a little is added (manipulation 3) until the adjustment is completed.

The thermometer is adjusted in a similar manner for the boiling-point method; by bringing it first into the vapor of the solvent, several cubic centimeters having been heated to boiling in a test-tube. A somewhat larger quantity of mercury is now removed from the thermometer proper, so that the thread of mercury comes to rest on the lower part of the scale, when the thermometer is reintroduced into the vapor of the solvent. Should too much mercury be thus removed from the thermometer, which can easily occur, it is returned most conveniently by means of the third manipulation.

Small drops of mercury which splash upon the walls of the reservoir and remain adhering to them, are shaken down into the mass of mercury remaining in the reservoir, by holding the thermometer upright and tapping it gently upon something beneath it, which is not too hard, say a book. On the contrary, a little drop of mercury in the upper funnel-shaped portion of the reservoir, protruding even into the capillary, is not thrown down in this manner, at least in the case of thermometers especially adapted to the freezing-point method. This is accomplished by tapping the side, as described above.

In this way the thermometer can generally be adjusted in a few minutes.

Trouble is encountered if the entire reservoir becomes filled with mercury, as sometimes happens in transportation. The attempt to unite the mercury in the reservoir with the mercury remaining in the bulb, by warming the latter, generally results in destroying the instrument. This is avoided by inverting the thermometer, fastening it in a clamp, and tapping it gently on something placed beneath it, as Beckmann[1] has suggested. A thread of mercury flows from the bulb through the capillary to the reservoir, an empty space remaining in the upper portion of the bulb. When the thread has united with the mercury in the reservoir, the thermometer is restored to its original position very carefully, so that the thread is not broken, and is then clamped in an upright position. The thread now siphons the mercury from the reservoir into the bulb. The siphoning can be interrupted at any moment desired, and the further adjustment completed according to the forgoing directions.

This artifice of inverting the thermometer to transfer mercury from the bulb to the reservoir can, of course, be employed in adjusting the thermometer, instead of warming the bulb.

A small, yet not inconsiderable inaccuracy results, as has been shown, from using the same thermometer

---

[1] E. Beckmann : Ztschr. phys. Chem., 15, 673 (1894).

as described, to carry out measurements at low and at high temperatures, for the reason that in the two cases the amount of mercury whose volume undergoes change, is different. If the thermometer is warmed from 0° to 1° the thread of mercury rises, in fact, about one degree on the scale. At higher temperatures less mercury is found in the bulb (the remainder in the reservoir), and, consequently, for an increase in temperature of 1°, the mercury thread would not rise one degree on the scale, but less than this amount. Fr. Grützmacher[1] has prepared a table, with which the degree as read on the thermometer can be corrected by calculation, by multiplying the degree as read by the coefficient given for the temperature.

Temperature.		A degree as read, when corrected, amounts to
—35—	—30	0.977°
0—	5	0.995
45—	50	1.015
95—	100	1.032
145—	150	1.045
195—	200	1.053
245—	250	1.055

The physical-technical Reichsanstalt in Charlottenburg will determine experimentally the value of the degree on a Beckmann differential thermometer, for twelve marks.

DETERMINATION OF MOLECULAR WEIGHT BY THE FREEZING-POINT METHOD

The molecular weight is derived by the freezing-point or cryoscopic method, from the lowering of the freezing-point which a solution shows with respect to

---

[1] Fr. Grützmacher: Ztschr. f. Instrumentenkunde, (1896) 202. Compare E. Beckmann: Ztschr. phys. Chem., 21, 252 (1896).

the pure solvent. This is made possible by the three following propositions, found experimentally, and which have led historically to the freezing-point method.

1. A solution freezes lower than the solvent.

2. The depression of the freezing-point is proportional to the concentration.

3. Equimolecular solutions in the same solvent, show an equal depression of the freezing-point.

The first of these has long been known. Solutions of sodium chloride and calcium chloride freeze only at temperatures which are lower than the freezing-temperature of water. The second principle was first discovered by Blagden in 1788, but was forgotten, and rediscovered by Rüdorff in 1861. Let the following numbers, obtained for a solution of potassium chloride in water, serve to illustrate it.

Potassium chloride. Per cent.	Lowering of the freezing-point.	Quotient of the two.
1	0.45°	0.45
2	0.90	0.45
4	1.80	0.45
10	4.40	0.44
12	5.35	0.45

The third principle was announced first by Coppet in 1871, as holding only for special classes of closely related substances. Its general validity was discovered by F. M. Raoult in 1882 to 1884, and, indeed, through the fortunate circumstance that he employed a large number of organic substances in his investigations. Certain irregularities, which generally complicated the relations with inorganic substance, did not appear in the case of organic compounds.

Raoult introduced the term "molecular depression" or "freezing-point constant" for values which can be compared. By molecular depression is to be understood the depression of the freezing-point of 100 grams of the solvent, by a gram-molecular weight (*e.g.*, 78 grams benzene) of substance dissolved in it.

In most cases it will not be at all possible to prepare a solution as concentrated as this, because the solubility of the substance to be investigated is too small, and further, because in concentrated solutions anomalies appear which would lead to erroneous results. But the molecular depression can be calculated from the depression of a dilute solution with the aid of principle 2. and, indeed, as follows:

In an experiment, a freezing-point lowering of D degrees was found for a solution of S grams of substance in L grams of solvent. Then a solution of 1 gram of the substance in the same amount of the solvent would produce a depression of $\dfrac{D}{S}$ grams, and a gram-molecular weight of the substance ($=$ m) the depression $\dfrac{MD}{S}$. If only one gram of the solvent is employed, the depression is $\dfrac{MDL}{S}$, and finally, for 100 grams of the solvent, $\dfrac{MDL}{100\,S}$. This is the molecular depression to be calculated, which we will designate by K :

$$K = \frac{MDL}{100 \; S}.$$

This value, from the third principle, is constant for a solvent in which any substance is dissolved. It is determined by experiment.

Example: 0.3999 gram of nitrobenzene in 19.91 grams of benzene, give a depression of 0.825°. The molecular weight of nitrobenzene is 123.

$$K = \frac{123 \times 0.825 \times 19.91}{100 \times 0.3999} = 50.5$$

A graphic extrapolation, which makes possible the calculation for infinite dilution, and thereby increases the accuracy of the value of K, will be discussed in the chapter which deals with the critical examination of the results.

As van't Hoff[1] has shown, the constant can be calculated from the absolute temperature T, at which the solvent freezes, and the latent heat of fusion w, from the following formula:

$$K = \frac{0.0198 \, T^2}{w}.$$

Example.—In the case of benzene, T $= 273 + 4.9$; w $= 29.1$.

$$K = \frac{0.0198 \times 277.9^2}{29.1} = 52.5.$$

A table containing a large number of solvents applicable to molecular weight determinations, with their corresponding molecular depressions, is given on page 106.

---

[1] J. H. van't Hoff: Ztschr. phys. Chem., **1**, 496 (1887).

If the value of the molecular depression of a solvent has been ascertained, by a series of determinations with substances of known molecular weights, it can then be employed for determining the molecular weights of substances whose molecular weight is unknown.

The depression is determined which the solution of a weighed amount of the substance produces in a weighed amount of the solvent. If this value is introduced into the above formula solved for m :

$$m = \frac{100 \, SK}{DL},$$

we have the molecular weight of the dissolved substance.

Example.—The analysis of a hydrocarbon obtained from coal-tar, showed that it corresponded to the simple formula C_5H_4. 0.5507 gram of the hydrocarbon in 18.65 grams of benzene gave a depression of the freezing-point of 1.170°.

$$m = \frac{100 \times 0.5507 \times 50.5}{1.170 \times 18.65} = 127.4.$$

The formula of the substance must, therefore, be doubled — $C_{10}H_8 = 128$. It is naphthalene.

If, as occasionally happens, the solvent acts chemically upon the substance, forming an addition-product, a formula adapted to the calculation of the molecular weight of the newly formed substance, can be obtained by slightly transforming the last formula.

If S grams of the substance take up 1 gram of the solvent, the weight of the unaltered solvent present is

L — 1, and in it are S + 1 grams. The molecular weight is then:

$$m = \frac{100\,(S+1)\,K}{D\,(L-1)}.$$

An example of this is furnished by arsenic trioxide. Every molecule when dissolved in water takes up three molecules of water.

$$As_2O_3 + 3H_2O = 2H_3AsO_3.$$

It should be remembered here, that no conclusion in reference to the probable addition of a solvent to a dissolved substance, can be drawn from the molecular weight determination of a dissolved substance in dilute solution. Such a conclusion is permissible only when the number of molecules is changed, as in the case of arsenic trioxide.

We often find in the literature the value C given, instead of the amount of substance and of solvent. C is the number of grams of the substance in 100 grams of the solvent.

$$C = \frac{100\,S}{L}.$$

The molecular weight is then calculated thus:

$$M = \frac{CK}{D}.$$

A number of authors, like Eyckmann,[1] introduced into the calculation the percentage composition[2] of

_____ ___ _

[1] J. F. Eyckmann: Ztschr. phys. Chem., 4, 497 (1889).

[2] The number of grams of the substance contained in 100 grams of the solution.

the solution, instead of the number of grams of the substance in 100 grams of the solvent. This has been recently abandoned, as was mentioned in the theoretical introduction to this section.

The freezing-point method was discovered by Raoult[1] in the years 1880–1885. It was first employed in the solution of chemical questions by Paterno and Nasini, and shortly afterwards by Ostwald, Hollmann, and others. Auwers[2] described an apparatus adapted to the chemical laboratory, and used it for determining the molecular weight of benzildioxime. Near the end of 1888, Beckmann[3] described an apparatus, which, since that time, has been almost exclusively used.

The theoretical foundation of the method, indicated in the introduction, had been furnished a year before by the fundamental work of van't Hoff,[4] on osmotic pressure.

The Simple Freezing-point Apparatus of Beckmann.— The Beckmann apparatus[5] is used for determining molecular weights by the freezing-point method. The solvent is placed in a thick-walled test-tube, A, which is about 2 cm. internal diameter. The substance whose molecular weight is to be determined

---

[1] F. M. Raoult : Ann. Chim. Phys. [5], **20**, 217 (1880) ; **28**, 133 (1883) ; [6] **2**, 66 ; 93, 99, 115 (1884) ; **4**, 401 (1885) ; **8**, 289, 317 (1886) ; Compt. rend., **102**, 1307 (1886).

[2] K. Auwers : Ber. d. chem. Ges., **21**, 701 (1888).

[3] E. Beckmann : Ztschr. phys. Chem., **2**, 638 (1888).

[4] J. H. van't Hoff : *Ibid.*, **1**, 481 (1887).

[5] E. Beckmann : *Ibid.*, **2**, 638 (1888) ; **7**, 324 (1891). The apparatus is furnished by the glass-blower, F. O. R. Götze, in Leipsic. Any desired change in the cover can be easily made by any tinner.

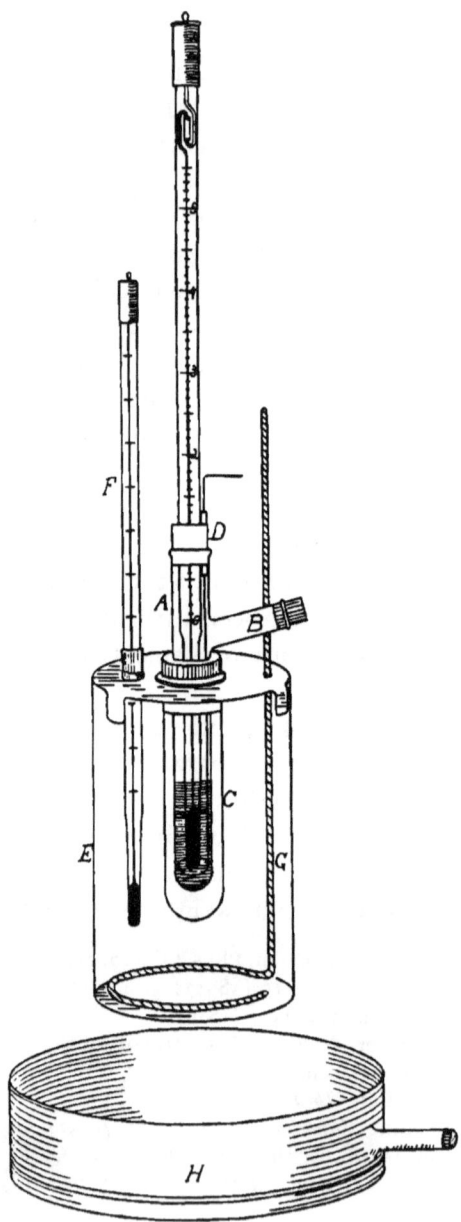

Fig. 13. Beckmann's freezing-point apparatus. One-fourth natural size.

is later introduced into the solvent in the test-tube, through the tube B, which is attached to the side of A. Both openings are closed with a cork; rubber[1] is not used. The thermometer is so introduced through the stopper in the top of the test-tube, that the mercury bulb is placed at a distance of about one centimeter from the bottom. A glass tube (D), whose edges have been fused off above and below, is introduced through a second, narrower hole. This glass tube serves as a guide for the stirrer[2] which is made by bending thick platinum wire. A piece of brass tubing is also employed instead of glass, because small particles of glass are easily torn off by the platinum stirrer, and then produce trouble. The stirrer may be placed directly through a hole in the cork.[3]

This freezing-vessel is surrounded with an air-jacket, which extends up to the side-tube. This is a shorter and somewhat wider test-tube, and is fastened to the freezing-vessel by means of a cork.

---

[1] A rubber stopper would not give the thermometer the necessary firmness.

[2] It is not allowable to use stirrers of any other metal, because such would produce trouble even with indifferent solvents. Glass stirrers are not to be recommended, independent of the fact that they must be made of fairly thick glass in order that they may not be too fragile, since they favor too strong undercooling. Glass tubes, with a platinum stirrer fused in below, are useful, — the so-called fusion glass is to be employed to produce a tight junction — but are always far inferior to the platinum stirrer. Compare F. W. Küster : Ztschr. phys. Chem., 13, 448 ; *Annk.* 1894. E. Beckmann : Ztschr. phys. Chem., 17, 108 ; *Annk.* 1895.

[3] The under ring of the stirrer should be only a little smaller than the diameter of the freezing-vessel, that it may have a clear path, and not catch under the thermometer, shake this violently, and be bent.

The freezing-vessel with its air-jacket is introduced into a thermostat, which is kept at a uniform temperature, and a few degrees below the point of solidification of the solvent. Generally a thick-walled glass vessel (e), is employed, such as is used in the Bunsen element, and holding about 1½ liters. It is closed above by a lid of thick brass or German silver, which is held firmly in position on the glass vessel by means of three strips bent downward. It is perforated in three places. In the center is a larger hole through which the freezing-vessel with air-jacket passes. It is advisable to introduce a gum or asbestos ring between the side of the air-jacket and the metal cover, to avoid breaking the side of the former. It is convenient to have three or four metal strips[1] fastened around the opening, and reaching downward. These hold the air-jacket upright, and with it the entire freezing apparatus proper. They also hold the air-jacket firmly in the freezing-liquid, when the freezing-vessel (A) is removed. A stirrer (g), made by bending strong German silver or nickel wire, is introduced through a second narrow opening. The third opening is provided with a tube, in which the thermometer (F), graduated to tenths of a degree, is fastened. The temperature of the liquid in the thermostat is measured by means of this thermometer.

The entire apparatus is conveniently placed in a tin vessel (H), in which any water, etc., which may overflow, is collected.

[1] E. Beckmann : Ztschr. phys. Chem., **7**, 325 (1891).

Carrying Out a Simple Molecular Weight Determination with the Beckmann Freezing-point Apparatus.

—In order to become familiar with the manipulation of the freezing-point apparatus, it is advisable at first to carry out a molecular weight determination under the simplest possible circumstances. For this purpose benzene is well adapted as a solvent, and any liquid, say an ethereal salt, can be used as the substance to be investigated.

The first problem is to adjust the thermometer. This is accomplished by following the mode of procedure given in the description of the thermometer. The problem is solved when, as previously indicated, the mercury thread comes to rest at a point about 0.2° to 0.5° from the upper end of the scale, the thermometer being immersed in partly solidified benzene. The same benzene is, of course, used for this purpose, as will be employed in carrying out the investigation. Otherwise it could easily happen, that in the actual experiment the solvent used might show a higher freezing-point than the less pure substance previously employed, and the top of the mercury thread come to rest above the scale.

Pure benzene is then weighed off in the carefully cleaned and dried freezing-vessel, in quantity sufficient to surround the bulb of the thermometer on all sides, when it is introduced, with a layer of about equal thickness. About 16 grams of benzene are necessary when the thermometer has a long bulb, while about 12 suffice when the bulb of the thermometer is short. The weighing should be made accurately to centigrams. The cork of the air-jacket, and then the jacket itself, are shoved up over the freezing-vessel, and the freezing apparatus, thus set up, is placed to one side. The vessel which is to contain the freezing-mixture is about three-fourths filled with pieces of ice, from the size of a walnut to that of a hazelnut, and this is then filled with water to within about 2 cm. of the top. The lid, with the stirrer and thermometer, is placed in position, and finally the freezing-vessel introduced. The vessel containing the freezing-mixture is wound around with a layer of felt, or with a folded towel, that the effect of the external temperature shall be made as small as possible. The temperature of this vessel remains constant from 0–1° if it is stirred now and then, and is therefore 4–5° below the freezing-point of benzene.

The next problem is to determine exactly at what point on the scale the mercury comes to rest when the benzene begins to solidify; i. e., the "determination of the freezing-point of the solvent." The benzene is cooled down by means of the freezing-mixture, and the mercury sinks slowly — slowly because the air-jacket prevents a too rapid equalization of temperature, — meanwhile a uniform temperature is secured throughout the benzene itself, by a gentle movement of the platinum stirrer, whose upper end is provided with a handle made by shoving a small cork over it. The marked effect of the stirrer is easily recognized by discontinuing the stirring for a minute or two. During this time the mercury usually falls only a little, but when the stirring is renewed, a rapid fall takes place, often amounting to several tenths of a degree. This is due to a mixing of the outer, much colder portions of the benzene, with the warmer portions lying nearer to the thermometer. It is important to stir even during the falling of the thermometer, in order that the colder layers of the benzene may be carried from the glass wall into the interior of the freezing-vessel. Otherwise, a layer of ice usually forms, before the thermometer has reached the real freezing-point. In this case fine needles of ice do not form, as is necessary, but the layer of ice grows thicker by further removal of heat. The thermometer does not show, then, the correct point of rest, and generally does not come to the correct position. In such cases, all the solid benzene which has separated should be immediately melted, and the determination of the freezing-point begun anew.

When the temperature is about one- to two-tenths of a degree below the freezing-point, already approximately determined in the adjustment of the thermometer ("undercooling"), vigorous stirring should be begun, *so that the stirrer strikes the bottom of the freezing-vessel* and thus starts the freezing process. In the case of benzene, the separation of the solid begins very easily under these conditions. The solid benzene separates in fine crystal plates, which are whirled about through the entire mass of liquid by means of the stirrer. These bring the liquid exactly to the true freezing-point, in that the warmer parts are cooled by the melting of the crystals, and the cooler parts warmed by the further formation of crystals. During the separation of the ice the thermometer rises rapidly, and its highest position gives exactly the freezing-point of benzene. It will remain at this point for a long time, provided the stirring is vigorous and continued.

It is important that the liquid should be somewhat undercooled before the formation of ice begins, in order that when this point is reached a large quantity of the crystals of the solvent should be formed at once. Otherwise, the thermometer would come to rest only very slowly, and not give an accurate result. This is accomplished as described, by stirring slowly at first, in order not to disturb the undercooling when it begins, and only when the undercooling has proceeded far enough, should the actual process of solidification be made to begin by vigorous stirring. A further advantage of the undercooling is this: The mercury of the thermometer comes to the true position more accurately[1] in rising than in falling from a higher temperature to the point to be read, as Pfaundler has shown. The error, thus arising, is due to the sticking of the metal in the narrow capillary. But it is limited in amount by the shaking to which the thermometer is subjected in stirring.

After the first reading is made, the freezing-vessel is removed from the air-jacket, which is left in the freezing-mixture, warmed gently with the hand until the solid benzene is just completely melted, and the determination is repeated several times. It is usually

[1] W. Ostwald: Hand- und Hilfsbuch zur Ausführung physico-chemischer Messungen, p. 53. (1893).

found that the first, also the second, and, indeed, sometimes the third determination, gives values which are
somewhat too low, because the thermometer, kept at
the temperature of the room, undergoes slow change,
until the elastic conditions of the glass, corresponding
to the low temperature, have been established. After
this, the remaining determinations of the freezing-
point give values which either agree exactly with one
another, or show deviations of only a few thousandths
of a degree. The deviations must not be greater than
0.005°. When the temperature of the experiment is
higher than that of the room, the deviations of the
first readings are generally somewhat too high.

To avoid these irregularities, it is advisable to keep
the thermometer, day and night, at the temperature
of the experiment,[1] when larger series of experiments
are to be carried out. This is especially desirable
when the temperature of the experiment differs widely
from the temperature of the room, as in the case of
naphthalene. This precaution is absolutely necessary
in those experiments which lay claim to greater
accuracy, and which are, therefore, carried out with
thermometers[2] whose scale is divided into thousandths
of a degree.

As nearly the same amount of undercooling as is

---

[1] K. Auwers: Ztschr. phys. Chem., 18, 596 (1895). R. Abegg
(*Ibid.*, 20, 216) regards this precaution as unnecessary for determinations at 0°, even when the greatest exactness is required, because the thermometer adjusts itself correctly after remaining a
quarter of an hour at the temperature of the experiment.

[2] E. H. Loomis: Wied. Ann., N. F., 51, 508 (1894). M. Wildermann : Ztschr. phys. Chem., 19, 91 (1896).

possible, should be secured in every repetition of a
freezing-point determination, because otherwise, in the
case of many solvents, c. g., naphthalene, constant
reading cannot be obtained. The amount of undercooling is observed, together with the freezing-point as
read. It is important to secure uniform undercooling,
especially in carrying out determinations of the
freezing-point of solutions.

The repetition of the determination of the freezing-
point can, further, be carried out in a somewhat different manner. A method which is to be recommended
for solvents which solidify with difficulty, is not
necessary in the case of benzene, because here solidification usually sets in easily, even with slight undercooling. The ice is not all melted, but *some crystals
are allowed to remain on the bottom.* Then the
freezing-vessel in the air-jacket is cooled down, and at
the same time only the upper portion of the liquid is
slowly stirred, in order that the crystals should not be
stirred up and melted. This is continued until the
thermometer shows an undercooling of from $0.1°$ to
$0.3°$. If, now, vigorous stirring is begun, the ice particles are whirled about and start the solidification
process.

At the same time let the third method be described
for starting solidification in the somewhat undercooled
liquid; i. e., "inoculation."[1] The inoculating apparatus necessary for this purpose is shown in Fig. 14.
About a cubic centimeter of the solvent is placed in

---

[1] E. Beckmann: Ztschr. phys. Chem., 2, 640 (1888); 7, 330
(1891).

the small test-tube. A narrow glass tube,
whose upper end is closed by a rubber tube
and glass rod, passes through a cork and dips
into this. By pressing upon the rubber tube
and then releasing it, some of the liquid is
made to enter the glass tube. The contents
of the test-tube are frozen by cooling, whereupon the liquid in the glass tube also freezes.
The rod of ice is loosened by gently warming
the tube, and is then shoved forward by
drawing the rubber tube somewhat further
over the glass tube. This is allowed to completely solidify again in the jacket, the little glass tube being drawn up somewhat in
the cork, so that the point of the rod of ice
no longer touches the remainder of the ice.

Fig. 14.
Inocula-
ting rod.

This inoculator is used when a liquid does not begin
to solidify when undercooled several-tenths of a degree,
as easily takes place with acetic acid and water. The
stirrer is then quickly raised from the undercooled liquid,
and the inoculator is introduced through the side tube
of the freezing-vessel, and made to touch the moist ring
of the stirrer, when the undercooled liquid clinging to
it, solidifies. The stirrer is allowed to drop quickly
before this ice melts, and the process of ice formation
is now brought about by vigorous stirring. The use
of this method, which serves as the last resource for
producing solidification, and is by no means to be
rejected, is generally unnecessary.

After the freezing-point of the solvent has been
determined by means of several concordant experi-

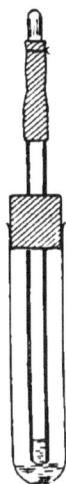

ments, the determination of the molecular weight proper can be proceeded with. In case the solubility of the substance will admit of it, several determinations are carried out with the same solvent, and as follows: A small amount of the substance is at first added to the solvent, and the depression of the freezing-point produced by it, observed, and then a second quantity is added, and so on.

The investigation of a liquid is the simplest. The liquid is drawn up through the narrow capillary, into a weighing pipette of the form given in Fig. 15, until the pipette is about three-fourths filled. The pipette thus filled, is weighed. It is convenient if, as shown in the figure, the pipette is even roughly graduated to cubic centimeters. To bring the substance into the freezing-apparatus, the capillary of the pipette is introduced into the side tube, and some of the substance dropped into the freezing-apparatus by blowing air through the rubber tube, which is drawn over the wider glass tube. It is better to introduce a small tube containing calcium chloride, between the mouth and the pipette. The amount of substance desired, from 0.03 to 0.1 gram, can be easily introduced by observing the graduations on the pipette, or by counting the number of drops which flow out. When a smaller amount of the solvent is taken (10–15 grams), and the substance to be investigated has a lower

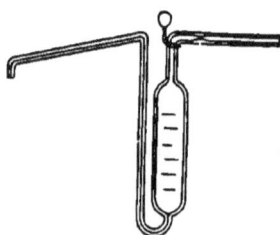

Fig 15.
Weighing pipette. One-fourth natural size.

molecular weight, a smaller quantity of the substance is used in the first experiment; when the conditions are the reverse, more substance is employed. Since the end of the capillary on the pipette is cut off obliquely, the liquid, when blown out, drops off cleanly, and the pipette can be withdrawn without loss and weighed again. The loss in weight gives the quantity of the substance used in the first experiment.

The freezing-point of the solution is determined in the same manner as that of the pure solvent. Three separate experiments suffice for this purpose. The depression should amount to from $0.05°$ to $0.2°$ in this first determination. If it is smaller, the determination is affected too seriously by errors of reading, of the thermometer, and from other sources. If it amounts to more than $0.2°$, we have worked with too great concentrations. It will be shown that the most accurate values can be obtained with dilute solutions.

After the first molecular weight determination has been made, an additional quantity of the substance is introduced into the freezing-apparatus, and its weight is added to that of the first quantity. The freezing-point of the more concentrated solution is then determined, and the total depression, of course, recorded. This is continued until a concentration of $5°-15°$ is reached, when the scale of the thermometer will no longer suffice.

Stress must be laid upon one point in the determinations. Care must be taken that the more concentrated solutions are not undercooled too strongly. The greater the undercooling, the larger the quantity of ice which separates at the moment of freezing, and therefore the greater the change in the concentration of the solution. A too great depression would be observed, and the molecular weight calculated, would be smaller than in the case of a determination carried out under normal conditions. The undercooling should not exceed 0.2° when benzene is used as a solvent.

The phenomenon that the mercury column remains at the true freezing-temperature for a shorter time the more concentrated the solution, and then begins to sink, is to be referred to the same cause. In dilute solutions, it will remain for a longer time at the highest point reached after the undercooling has been removed by freezing. The mass of the frozen solvent increases with the time during which the freezing-vessel remains in the freezing-mixture, the concentration increases, and therefore the freezing-point sinks.

On the other hand, finely divided ice separates more easily from concentrated solutions than from dilute.[1] Every ice particle is surrounded, when formed, with a layer of concentrated solution, which prevents its further growth. A small quantity of fine particles, which therefore have a large surface, suffice to keep the freezing-temperature constant.

Finally, attention should be called to the fact that the thermometer is not adjusted with the same degree of exactness and certainty, in experiments with concentrated solutions as with dilute, since, as observed above in the former case, changes in the concentration

---

[1] F. W. Küster: Ztschr. phys. Chem., 13, 448 (1894).

have a marked influence. But since in these experiments the total depression is very large — generally
several degrees — an error of even 0.01° does not mate-
rially affect the result. The limit of error in normal
determinations is as high as 5 per cent. A numerical example will serve to elucidate this description.

$$C = 50.$$

ETHYL ACETATE IN BENZENE. CRYOSCOPIC METHOD.

In 15.07 grams benzene gave :

Grams ethyl acetate	0.0482	0.1369	0.2703	0.4508	0.7176	0.8920
Depression	0.167°	0.490°	1.077°	1.754°	2.638°	3.266°
Molecular weight	95.7	92.7	83.3	85.2	90.3	90.6

The molecular weight, calculated from the formula
$CH_3COOC_2H_5$, is 88.

Mechanical Stirring Device. — The stirring, in
carrying out a molecular weight determination by the
freezing-point method, is a very troublesome operation, as would be inferred from the foregoing description. Therefore, a stirring device driven by an
electromotor or a water turbine, is employed, when a
series of determinations is to be carried out, or when
molecular weight determinations are to be frequently
made, as is the case in a scientific laboratory. A
mechanism[1] to be driven by water, which has been
found to be efficient in a large number of determina-

---

[1] The mechanism is beautifully made by the mechanic,
H. Wittig, in Greifswald, for from 12 to 15 marks (without
turbine). In ordering an apparatus for a turbine already at
hand, it is well to forward this. The turbine can be easily removed
at any time and used for other purposes.

tions, under very widely different conditions, is shown in Fig. 16.

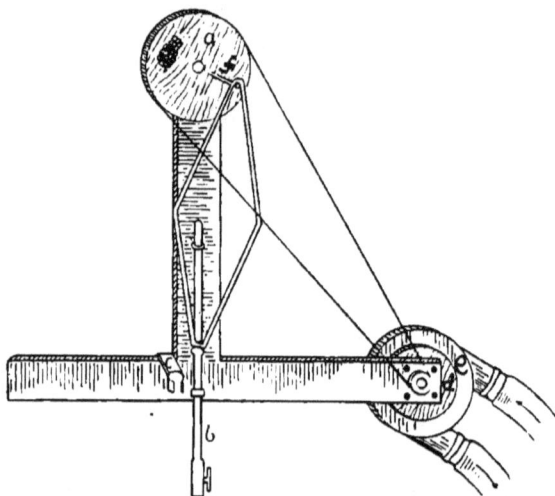

Fig. 16.
Mechanical stirring device.   One-sixth natural size.

Two strips of zinc or brass are riveted together at right angles, as shown in Fig. 16.   A round wooden disk (d), $\frac{1}{2}$ cm. thick, is fastened to the right end of the horizontal strip, with four screws.   The disk, as well as the metal strip, is provided with a conical hole.

The projection from a Rabe's[1] turbine which is also conical externally, fits into this, so that the driving wheel protrudes even beyond the metal plate.   The turbine drives with a cord, an easily movable disk of hard wood (a), which, in turn, by means of a rhomb, moves a driving-rod (b) up and down, in two guides.   The length of the stroke is regulated by shoving the

[1] H. Rabe: Ber. d. chem. Ges., 21, 1200 (1888).

upper attachment of the rhomb along the slit (f), of
the upper disk, so that the stirrer of the freezing-
apparatus, during the stirring, is not lifted out of the
liquid. A circularly bent clamp, open in front, serves
to hold the thermometer, and is so attached to the
strip that the thermometer, when shoved in the clamp
and held firmly by it, stands close to the upright strip,
while the stirrer, attached to the lower end of the
driving-rod (b), is moved as desired. The velocity can
be either increased or decreased by regulating the
flow of water. The mechanism allows the freezing-
apparatus to be easily removed from the stirrer, as is
always necessary before every new determination, that
the freezing-vessel may be removed for the purpose of
thawing out its contents, the thermometer being removed from the clamp. and the platinum stirrer from
the ear of the driving-rod.

The stirring device is held by two ordinary firm
iron stands, such as are used in the laboratory, heavy
weights being placed upon their bases. The wide
open clamp of one of them seizes the wooden disk
(d), to which the turbine is attached, above and below;
the clamp of the other seizes the left end of the horizontal zinc bar in front and behind. The height is
regulated by shoving the two clamps up and down on
the stands. The stirrer must be placed so that it
agitates the entire mass of the liquid.

Any one who has experienced the great saving of
labor, which results from working with such a stirring
device, will not willingly be without it.

If it should be necessary to clamp the stirring

device very high upon the iron stands, it is recommended not to support the thermometer by means of the clamp attached to the stirring device, but by a similar clamp attached to a third stand. The oscillations which are produced by a high, and therefore less securely held, stirring device, are thus prevented from being transmitted to the thermometer, and interfering with the reliability of the reading.

Procedure when Hygroscopic Solvents are Employed.— A modification of the simple freezing-point method is required, when molecular weight determinations are to be carried out with hygroscopic solvents. Some air enters the apparatus every time the stirrer is raised, and escapes from it whenever the stirrer is lowered. The amount of air which enters the apparatus is considerable, when a stirrer with glass handle is employed, because of its larger diameter. The air which enters is well saturated with water, because the cooling-jacket gives off water-vapor. The result is that, from determination to determination, the amount of water which enters the freezing-vessel and mixes with its contents, is so considerable that the freezing-point continually sinks, and, indeed, from one determination to another, from $0.005°$ to $0.02°$. If there are 10 grams of glacial acetic acid in the freezing-vessel, then, on account of the small molecular weight of water, the freezing-point is depressed about $0.02°$ for every milligram of water which the acetic acid has taken up.

The following values were obtained in the experiment to determine the freezing-point of acetic acid, using the apparatus described and a platinum stirrer.

4.767	4.698	4.656
4.746	4.695	4.638
4.720	4.676	4.620

This source of error can be completely avoided by means of an arrangement devised by Beckmann.[1] A somewhat wider freezing-vessel, with an internal diameter of 2.4 cm., is employed. It is widened at the upper end to 2.7 cm. The entire length is 22 cm. The device given in Fig. 17, which represents a somewhat modified guide of the stirrer, is held in position, together with the thermometer, by means of a cork, and the former is attached to the thermometer by a ligature, a piece of cork being placed between. The figure shows at once how a wire bent at right angles, and which can enter the ear of the mechanical

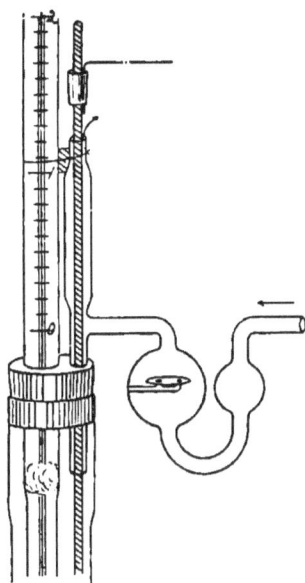

Fig. 17.

Introduction of the stirrer when hygroscopic substances are used.
One-third natural size.

stirring device, is fastened above with a small cork, to the stirrer with a glass handle. A glass tube opens into the lower part of the side of the guide. Through this is introduced a current of air, which is furnished by a water-blast, or from a gas-generator (CO_2), and is

[1] E. Beckmann : Ztschr. phys. Chem., 7, 324 (1891).

dried in wash-bottles filled with concentrated sulphuric acid, or in a drying tower. The air is freed
from the last traces of moisture, which might come
from the long rubber tube which conducts the current
of air to the apparatus, by passing it through the two-
bulbed drying device, containing a few drops of concentrated sulphuric acid. To prevent a spattering of the
sulphuric acid in the bulb, into the tube through
which the stirrer moves, a mica plate is placed in the
bulb next to the stirrer, on a glass rod fused into the
side of the bulb. The dry air escapes from the upper
end of the guide, as shown by the arrow; some little
dry air escaping when the stirrer is raised, but no
moist air can enter from without. The velocity of
the current of air is so regulated, that the bubbles just
cannot be counted. The current of air is not interrupted during the entire series of experiments.

This device works very satisfactorily. The freezing-
point of glacial acetic acid remained constant for 2 ½
hours, giving the values: 3.968; 3.970; 3.968. The
following values were found for phenol within a half-
hour. Without this arrangement: 4.498; 4.490; 4.482;
4.465. With this arrangement: 4.436; 4.430; 4.430;
4.436.

If the Beckmann device described is not available,
the following can be used: A current of dry air can be
led in through the cork closing the side tube of the
freezing-apparatus. This flows through the upper
portion of the freezing-tube, and escapes between the
stirrer and its ordinary guide. It must be taken into
account that the air current carries off some vapor of

the solvent from the upper portion of the freezing-vessel, but this source of error is certainly not very considerable. At all events, satisfactory molecular weight determinations can be made by this means.

A little moist air enters the freezing-vessel by both methods, during the introduction of the substance. The resulting error is, however, too slight to be taken into account in practice in the chemical laboratory.

Beckmann[1] has subsequently recommended another contrivance for keeping out the moisture. The freezing-vessel is completely closed. A stirrer, to which a wrought iron ring covered with platinum is attached above, is placed in it. This can be attracted by an electromagnet, which is attached to the outside of the freezing-vessel, and thus raised. It falls again on breaking the current. The alternate making and breaking of the current is effected by a Mälzel's metronome, modified for this purpose. A Gülcher thermopile serves to furnish the current.

The original contrivance has recently been modified by Beckmann,[2] thus: The stirrer enters the freezing-vessel through a mercury valve, which prevents the entrance of air and moisture during the experiment. The contrivance sketched in the place above cited, one can easily make for himself out of glass tubing and corks. Formic acid, acetic acid, succinic acid anhydride, cresol, phenol, are to be treated as hygroscopic substances.

---

[1] E. Beckmann : Ztschr. phys. Chem., **21**, 239 (1896).
[2] E. Beckmann : *Ibid.*, **22**, 616 (1897).

Determination of the Molecular Weight of Solids.

—Solids are introduced into the freezing-apparatus either in pieces or as powder. Pieces of suitable size are split off of crystals and larger coherent masses, with a knife; the projecting particles, which could be easily broken off, are removed, the pieces weighed, and used for the experiment. Finely crystalline and pulverulent substances are pressed into tablets, using the tablet press,[1] shown in cross-section in Fig. 18. The plug (*a*), with the concave side upward, is shoved down to the bottom of the thick-walled steel tube, polished smoothly inside and out, and having a diameter of about 9 mm. The roughly weighed powder is then poured in, the second plug (*b*), with the concave side downward, and then the rod (*c*), which receives the blows from the hammer, introduced. The tablet press thus arranged, is placed upon a piece of solid wood, the rod (*c*) struck several blows with a wooden hammer, while the apparatus is pressed firmly upon the support with the left hand, to prevent the lowest plug from being driven out. When the tablet is considered to be sufficiently coherent — which varies with the substance used — the tube is placed horizontally upon the table, and the lowest plug and then the tablet are driven out of the tube, by gently striking the rod (*c*). If an edge projects it is removed by a knife. Such substances as benzoic acid, naphthalene, and borneol, can be easily compressed into

Fig. 18.
Tablet press.
One-third
natural size.

[1] E. Beckmann : Ztschr. phys. Chem., **4**, 548 (1889).

tablets, while such as the anhydride of arsenious acid are compressed only with difficulty, using violent blows.

The less dense the tablet the more easily it dissolves, but, on the other hand, it is more fragile. The tablets are not made heavier than 0.3 gram with substances of medium density. If it is desired to introduce more substance at once into the apparatus, two or more tablets are used.

The press, after it is cleaned, is placed in a desiccator containing some concentrated sulphuric acid, in which it is kept protected from moisture. If it should not be used for a long time, it should be covered with a thin layer of vaseline. The formation of rust, which would make the press useless, is thus avoided.

Another form of tablet press has been described by Gerhard,[1] in which the powder is not pressed into tablets by hammering, but by means of a screw. This press has the advantage that the mould can be made of different materials, such as porcelain and ivory.

Substances which are very difficultly soluble, can be weighed as powder in small vessels made of fine platinum gauze. Generally, however, loosely pressed tablets suffice.

When it is desired to introduce the tablets into the freezing-apparatus, the handle of the stirrer is so turned that it does not stand in front of the side tube of the freezing-vessel. The tablet is then introduced, and when it is large, the thermometer is shoved a little to one side. The tablet, when it lies on the bottom of the vessel,—or when lighter than the

---

[1] V. Gerhard : Ztschr. phys. Chem., 15, 671 (1894).

solvent is held by the stirrer, — is dissolved by warming the bottom of the vessel very carefully over a flame. The process of solution is hastened by gently stirring. Some precaution is necessary, here, to prevent the drop of mercury which enters the reservoir of the thermometer from being shaken off. It is better to raise the thermometer so that its bulb does not dip into the liquid. When the substance is dissolved, the solution is cooled by dipping the freezing-vessel, without air-jacket, into water, etc., until the mercury comes down on the scale. The vessel is then dried upon the outside, and the determination of the freezing-point proceeded with as usual.

The Thermostat. — The purpose of the thermostat is to surround the freezing-apparatus proper, with a medium whose temperature remains constant at about $3°-4°$ below the freezing-point of the solution. Benzene is the most convenient solvent to use, since the required temperature can be easily secured in the thermostat, and maintained constant for a long time by filling it with melting ice. The maintenance of other temperatures requires special attention to the existing temperature, and special practice in temperature regulation. The glass vessel of the thermostat is surrounded with a layer of felt, or with a folded towel, to reduce the effect of external temperature to a minimum, or for higher temperatures it is wound around with several layers of asbestos paper.

Water is to be taken into account as a solvent which freezes lower than benzene. It generally suffices to fill the cooling-vessel with pieces of ice

about the size of a hazel-nut, and to throw in about a
handful of sodium chloride, so that the temperature of
the mixture is about $-4°$ to $-5°$. The brine is to be
pipetted off from time to time, and ice and salt added
as needed. The temperature of the cooling-bath
must be observed in this case as in the following, in
every determination of the freezing-point. Care must
be taken that the temperature does not vary more than
a few tenths (0.4°–0.6°) of a degree. Larger variations affect the freezing-point.

It is very convenient to use a cryohydrate, formed
of finely broken ice[1] and a large quantity of potassium
nitrate. This gives a constant temperature of 2.7°.

When solvents are employed which solidify between
5° and 15°, the cooling cylinder is filled with water,
and pieces of ice added as required. The temperature
of the mass is frequently made uniform by stirring.
When too much water collects in the cooling vessel,
the excess is removed by means of a pipette with a
wide opening below, by inserting it in the tube in
which the thermometer is generally placed. A temperature, to within a half degree, can be readily
maintained in this manner, with some practice, without interfering with the usual process of the molecular
weight determination.

If a temperature of 15°–25° should be maintained,
some water is removed from the thermostat from time

---

[1] A large mortar of hard wood, with a lid of sheet zinc perforated in the middle, and a heavy pestle, can be used for breaking
up the ice. W. Ostwald : Hand- und Hilfsbuch zur Ausführung
physico-chemischer Messungen, page 60 (1893).

to time with a pipette, and warm water added from a wash-bottle conveniently at hand.

When a temperature of from 25°–50° is to be maintained in the bath, it is readily secured by means of a small steam heater. A lead tube of 6 mm. external, and 3 mm. internal diameter, is introduced into the glass vessel of the thermostat, through an opening in the cover. It is coiled twice upon the bottom, and then returned through the same opening in the cover. The tube is placed at some distance from the walls of the vessel, that the stirrer can be moved up and down close to the wall, without interference. It is not difficult to maintain a temperature constant to within a few tenths of a degree, by regulating the supply of steam, shutting it off entirely from time to time, and when the temperature begins to fall, turning it on again.

If the temperature of the thermostat should be still higher, a large beaker, on which the cover of the thermostat is fitted, is employed. It is surrounded with several layers of asbestos paper, and heated on an asbestos board with a small Fletscher burner. The beaker is suspended in the ring of a laboratory stand, to prevent it from falling. A string should be wound around the stirrer in several places, to prevent it from breaking the thin-walled beaker. In this case also, it is not difficult with a little practice, to maintain a constant temperature. The slight movements of the thermometer, graduated to tenths of a degree, indicate how it should be regulated. The flame should be so adjusted that the thermostat should tend to have a

temperature a little higher than that desired, and the
temperature should then be lowered, from time to time,
by introducing a little cold water from a pipette.

In order that the conditions throughout a series of
experiments should be as nearly as possible the same,
the temperature of the thermostat should be lowered
by the same amount as the freezing-point of the solution. This point has not, in general, been sufficiently
regarded. But it is clear that when a constant temperature is maintained in the thermostat, the difference
in temperature between the freezing-vessel and the
thermostat becomes less, with increasing concentration
of the solution, and the consequent lowering of its
freezing-point. The errors arising from a disregard of
this point, are not to be neglected with strong solutions.

A large water-bath, whose temperature is kept constant by an automatic regulation of the supply of
heat, should, consequently, not be chosen. The
devices already described, which make it possible to
maintain any desired temperature, are to be preferred.
It would be better to regulate the temperature of the
thermostat in this way, also in the experiment
described above with benzene. It was not recommended in that place, that the carrying out of a
practical example might be made as easy as possible
for the beginner.

The Behavior of Individual Solvents.— All stable
substances which dissolve[1] the substance to be investi-

---

[1] In order to test whether the substance to be investigated is
sufficiently soluble in the solvent chosen, the attempt should be
made to dissolve some small particles of it in a few drops of the
solvent, in a small test-tube of about 1 cm. diameter. The solution

gated in sufficient quantity without acting on it chemically, and whose freezing-point can be reached without too great inconvenience, can be used as solvents for determining molecular weights by the freezing-point method. Of the substances whose melting-point lies considerably above 100°, only a few metals have thus far been employed. The substances used up to the present, are given in the following table, together with the melting-point, the latent heat of fusion, and the molecular depression of each.

CRYOSCOPIC SOLVENTS.

Solvent.	Melting-point.	Latent heat of fusion.[1]	Molecular depression.
Acetophenone[2]	20°	30	56.5 (?)
Acetoxime	59°	41	52.9
Ethylene bromide[3] ..	9°	13	117.9
Formic acid	8°	56	27.7
Anethol	21°	28	61.2
Aniline	6°	24	58.7
Azobenzene	68°	30	77.6
Benzoic acid	122°	39	78.5
Benzene	5.4°	31	50
Benzophenone	48°	23	87.8
Succinic acid anhydride	120°	48	63
Bromoform	8°	12	133
p-Bromphenol	63°	23	98

should then be cooled until the solvent freezes, when it should be observed whether the substance separates. It is urgently advised to make this test in every case, lest the experiment fail, because the cold solvent is not capable of dissolving the substance in sufficient quantity.

[1] The latent heats of fusion are calculated from the molecular depressions, by the van't Hoff formula.

[2] Pure acetophenone is difficult to obtain. E. Beckmann : Ztschr. phys. Chem., 17, 110 (1895).

[3] Ethylene bromide should be kept in the dark.

Solvent.	Melting-point.	Latent heat of fusion.	Molecular depression.
p-Bromtoluene	28°	22	82.2
Cetyl alcohol	49°	34	59.7
Chloral alcoholate	46°	27	74.4
Diisoamylsulphone	31°	15	119.0
Diisoamylsulphoxide	37°	36	53
Diisobutylsulphone	17°	26	63
Diisobutylsulphoxide	68°	34	67
Dimethylaniline	0°	26	58.0
m-Dinitrobenzene	90°	27	98.0
Diphenyl	70°	29	79.4
Diphenylamine	54°	23	88
Diphenylmethane	26	27	65.6
Acetic acid	17°	44	39
Isoapiole	55°	27	80
Caprinic acid	31°	41	44.7
p-Cresole	36°	27	69.6
Crotonic acid	72°	39	61
Laurinic acid	44°	45	44
Methyl oxalate	54°	40	53
Naphthalene	80°	36	69
α-Naphthylamine	50°	26	78
Nitrobenzene	3°	21	70.7
Palmitic acid	62°	51	44
Phenanthrene	99°	23	120
Phenol	40°	27	72
Phenyldisulphide	61°	32	69
Phenylpropionic acid	49°	25	82.6
Phosphorus	44°	5	384
Resorcine	110°	45	65
Stearine (commercial	56°	44	49.2
Stearic acid (commercial)	53°	49	42.5
Nitrogen dioxide (N_2O_4)	-10°	33	41
Sulphobenzid	128°	26	122
Sulphonal	126°	36	87

Solvent.	Melting-point.	Latent heat of fusion.	Molecular depression.
p-Toluidine[1]	45°	36	56.2
Thymol	51°	28	73.9
Urethane (ethylate of carbamic acid)	49°	41	⁻49.6
Urethylane (methylate of carbamic acid)	52°	49	43
Water	0°	80	18.5
p-Xylene	15°	38	43

These solvents are to be employed only when perfectly pure. In the cases of stearine and stearic acid, the commercially pure product suffices. Substances which are easily decomposed must be purified just before using. Some statements should be made in reference to the use of the most important of these solvents.

Formic and *acetic acids* behave very similarly. Both of these easily show strong undercooling, even with violent stirring. They, therefore, generally require inoculation, either with the inoculating rod, or with some ice particles which are left unmelted. Since they are very hygroscopic, it is necessary to conduct dry air in through the stirrer guide. The specimen to be used in the experiment must be carefully freed from water. This is most easily accomplished with acetic acid, by starting with several kilograms of the acid, partly freezing it, pouring off the unfrozen portion, melting the solid formed, and allowing it to partly freeze again. After removing again the unsolidified portion, an acid remains which can be used. The same pro-

---

[1] p-Toluidine must be carefully dehydrated. Compare M. Stephani : Disst., Zürich (1896), p. 25.

cedure is adopted with formic acid, if a sufficient quantity is available.

When *water* is to be used, ordinary distilled water suffices. It is generally necessary to make it freeze by inoculation.

It is necessary to vigorously stir *naphthalene, phenol, azobenzene,* and *stearic acid,* near the point of solidification, and to wait until the mercury has reached the highest point, which usually requires considerable time. To obtain a sufficient undercooling the liquid is stirred very slowly, until it is about $0.1°$ below its freezing-point, and then stirred vigorously. It is important that the entire mass of the liquid should be agitated.

Stearic acid recommended by Eyckmann,[1] proved to be useful in some experiments in which the Beckmann apparatus was employed. When it is used, the temperature of the cooling-bath must be kept six or seven degrees below the melting-point of the substance used, which must be previously ascertained, and a greater undercooling of, indeed, $0.5°$ to $0.6°$ must be effected. The adjustment of the thermometer for stearic acid requires some attention.

Nitrobenzene is not so well adapted to more exact determinations, because constant values for the adjustment cannot be easily obtained with it.

Azobenzene behaves like naphthalene. It presents this advantage, that it has a smaller tendency to sublime. On the other hand, the determination of a

---

[1] J. F. Eyckmann : Ztschr. phys. Chem., **4,** 500 (1889).

constant freezing-point is more difficult. It is also necessary at times to prevent too strong undercooling by inoculation.

Naphthalene and *phenol* easily sublime into the upper part of the freezing-vessel, and can be returned to the lower portion only, in part, by melting. In the later experiments of a longer series, a small quantity (0.1 to 0.3 gram) is to be subtracted from the amount of the solvent to be employed in the calculation, as a correction. It is especially necessary with naphthalene to have the same undercooling in every determination.

As regards the amount of *undercooling*, it should be greater for solvents with a large heat of fusion, (about 0.3°–0.5°), since otherwise so little ice would separate, that when mixed with the remaining mass of the solvent, it would not suffice to bring this up to the true freezing-temperature. It is especially desirable to have considerable undercooling (about 0.5°) when formic acid, acetic acid, and water[1] are used.

Solvents which have a very small heat of fusion, as nitrobenzene and phosphorus, show freezing-points which sometimes vary some hundredths of a degree. This is due to the fact that their crystals effect only small changes in temperature, when, on mixing with the solution, they increase in size or melt. But since the molecular depression of these substances (especially phosphorus) is very considerable, the results can be used for a molecular weight determination.

---

[1] M. Wildermann recommends an undercooling of at least 0.6° to 0.7° with water. Ztschr. phys. Chem., **15**, 359 (1894).

Acetic acid,[1] which has been strongly recommended by V. Meyer and Auwers, is admirably adapted for laboratory practice; first, because it can be worked with conveniently, and second, as will be shown later, considerable irregularities are scarcely ever found in the results. Further, *benzene* is well adapted to this purpose, and can be worked with more conveniently than acetic acid, and although it sometimes gives irregular results, nevertheless it is very useful in determining molecular values when the explanations to be given later (chapter containing results) are taken into account. It is all the more useful, since the large value of the constant diminishes the effect on the result of small errors in the measurement of temperature. *Naphthalene*, which in other respects is as good for this purpose as benzene, at its melting-point, 80°, is an excellent solvent for many substances which are only slightly soluble in bezene at 5°. Finally *phenol* is to be mentioned. This, like acetic acid, gives results which are simpler in their relations.

Solvents Which Cannot Be Used in Certain Cases.—Solvents which are isomorphous with the substance to be dissolved in them, or which are closely related to it chemically, cannot be used for molecular weight determinations. These solvents do not separate as pure solids, but there separates a mixture of the solid solvent and the dissolved substance, which

---

[1] K. Auwers : Ber. d. chem. Ges., **21**, 707 (1888). K. Auwers and V. Meyer : *Ibid.*, **21**, 1068 (1888). E. Beckmann : Ztschr. phys. Chem., **2**, 742 (1888).

is to be regarded as a solid solution.[1] In these cases, a smaller quantity of the dissolved substance remains in the solution than was added to the solvent. The depressions of the freezing-point found, are, therefore, too small, and the molecular weights calculated are too large.

This irregularity is shown by : Solutions in *benzene*, of thiophene, pyrrol, pyrroline, pyridine, pyrroleine, piperidine, quinoline, tetrahydroquinoline.

Solutions in *naphthalene*, of indol, indene, quinolene, isoquinolene, tetrahydroquinoline, β-naphthol, pyrrol, pyrrolene.

Solutions in *phenanthrene*, of carbazol, anthracene, acridine, tetrahydrocarbazol, diphenylene oxide, naphthoquinolene.

Solutions in *diphenyl*, of dipyridil, tetrahydrodiphenyl.

Finally, solution in *benzoic acid* of salicylic acid, α-carbopyrrolic acid, in *succinic acid anhydride*, of maleic acid anhydride, and in *acetophenone*, of acetylpyrrol, acetothienone.

This irregularity is not found for substances with open chains; *e. g.*, butyric acid in *crotonic acid*, oleic acid in *stearic acid*, give normal values.

The presence of side chains either in the solvent or in the dissolved substance, tends to give normal values. Pyrrol and thiophene dissolved in *p-xylene*, and, further, methyl pyrrol dissolved in *benzene*, give normal depressions.

---

[1] J. H. van't Hoff : Ztschr. phys. Chem., **5**, 322, 334 (1890).

It was possible in some of these cases, to prove directly that some of the dissolved substance had separated on freezing. The solid which had separated was isolated and analyzed, the amount of the dissolved substance present in the mother-liquor included in the crystals, being determined by a peculiar device,[1] and left out of account.

Iodine dissolved in benzene showed the same peculiarity as the above-mentioned ring compounds. Its behavior has been shown very clearly, through an investigation by Beckmann[2] and Stock, by studying it, on the one hand, in other solvents, which allowed no iodine to crystallize out with them, and on the other, by determining the amount of iodine in the solid benzene which separated out.

Antimony[3] dissolved in tin, behaves just as irregularly.

Increase in Accuracy in Investigating Very Dilute Solutions.— In studying scientifically the law which lies at the foundation of the freezing-point method, pains has been taken to refine the method, and thus to increase, as far as possible, the accuracy of the results. Since the laws of osmotic pressure appear clearly, only in very dilute solutions, this investigation must extend to such solutions, and to the very small depressions of the freezing-point given by

---

[1] A. V. Bijlert : Ztschr. phys. Chem., **8**, 343 (1891). Compare also E. Beckmann: *Ibid.*, **27**, 609 (1897).

[2] E. Beckmann and A. Stock: *Ibid.*, **17**, 107 (1895).

[3] C. T. Heycock and E. H. Neville : Chem. News, **59**, 157 (1889).

them. These experiments have partly accomplished
their purpose. Numerous sources of error have been
discovered, but others are not yet sufficiently understood to enable one to make a proper correction for
them.

Much larger volumes of liquid (about 1 liter[1]) are
used for exact measurements. The surface is then
small in proportion to the mass, and variations in the
external temperature have a smaller influence. The
pure solvent is poured at first into the apparatus,—
water alone having been used up to the present,—and
the freezing-point of the water ascertained. A definite
volume of the solvent is then removed by means of a
pipette, and an equal volume of a standardized solution of the substance to be investigated, added. A
solution of known concentration is thus conveniently
prepared in the apparatus.

A plate, with openings cut opposite one another,
serves as a stirrer. It is placed horizontal, and moved
up and down by means of a rod. A vigorous movement of the liquid particles is effected by a special
arrangement[2] of the sections. Hitherto stirrers of
porcelain, silver, and brass have been used. The use
of a platinum stirrer appears desirable also in this case,
and all the more so, since working with a platinum
stirrer is not so expensive as we generally think, since
the larger platinum words—*e. g.*, Heraeus in Hanau

---

[1] H. C. Jones: Ztschr. phys. Chem.. **11**, 111, 529 : **12**, 623
(1893); P. B. Lewis : *Ibid.*, **15**, 367 (1894).

[2] H. C. Jones: *Ibid.*, **11**, 533 (1893); P. B.Lewis : *Ibid.*,
15, 368 (1894).

—are generally ready to take back platinum apparatus furnished for scientific investigations, after the investigation is finished.   It is therefore necessary to
pay only for making the apparatus.

The velocity of the stirrer has a marked influence,
according to the experiments of Nernst and Abegg,[1]
since heat is continually being produced in the liquid by
stirring, and this raises the real freezing-point.   Consequently, they employed a device for driving the
stirrer, whose speed can be accurately regulated.   The
effect of the stirrer is of less significance in ordinary
molecular weight determinations, in which the liquid
has relatively a larger surface.

Thermometers[2] have been occasionally employed
whose scale was graduated to thousandths of a degree,
and which, accordingly, had a range of from only $\frac{1}{2}°$
to $1°$.   Also, thermometers graduated to hundredths,
have been read with a microscope[3] with ocular microm-
eter, which made it possible to estimate the ten-
thousandths of a degree.   Readings of equal accuracy
were possible by these two devices, provided the
graduation on the thermometer divided into hundredths, is made fine enough.   It is, however, a question whether the use of a thermometer graduated to
thousandths is not to be abandoned, because the bulb
required is too large.   The possibility of heating a
mass of about 300 grams of mercury, with sufficient

---

[1] W. Nernst and R. Abegg : Ztschr. phys. Chem., 15, 687
(1894) ; R. Abegg: Ibid., 20, 212 (1896).

[2] H. C. Jones: Ibid., 11, 110 (1893) ; P. B. Lewis: Ibid.,
15, 366 (1894).

[3] E. H. Loomis: Wied. Ann., N. F., 51, 506 (1894).

uniformity with the arrangement used, appears to be doubtful. At all events, the sources of error inherent in the method of work, are, up to the present, of more consequence than the errors due to the inexact reading of the temperature.

The influence of the cooling-vessel, whose temperature must be kept very constant, is of the very greatest significance. Nernst and Abegg[1] surrounded the freezing-vessel on all sides, also above, with a freezing-mixture which was contained in a vessel protected from the temperature of the room by a thick layer of felt. The temperature of the room should be as near as possible to the freezing-temperature. A correction for the effect of the temperature of the freezing-bath, and for the action of the stirrer, obtained from observations, was first applied by Nernst,[2] through a method of calculation which still needs some modification. This method was discussed by Wildermann,[3] who endeavored to so arrange the conditions of his experiments, that corrections could be done away with. Raoult[4] sought to solve the problem in a simple manner, which was the more desirable, since the values of Nernst's correction cannot be determined with any great degree of certainty. His method is, moreover, very closely related to the procedure of Wildermann above referred to.

[1] W. Nernst and R. Abegg : Ztschr. phys. Chem., 15, 685 (1894) ; R. Abegg; Ibid., 20, 211 (1896).
[2] W. Nernst and R. Abegg : Ibid., 15, 682 (1894).
[3] M. Wildermann : Ibid., 19, 63 (1896).
[4] F. M. Raoult : Ibid., 20, 601 (1896).

CRITICAL EXAMINATION OF THE RESULTS

Smaller Anomalies Inherent in the Method. — If a series of molecular weight determinations is carried out, with increasing concentration of the same substance in the same solvent, it would be expected that all of the determinations would give the same value. But such is not the case. The results do not vary irregularly about a mean value, but show a quite regular decrease or increase. The smallest values are generally found at the greatest dilution. The molecular weight found, increases with increasing concentration. This particular kind of increase is shown by the following values for the molecular weight of naphthalene (m = 128) in benzene, published by Beckmann.[1]

Concentration.	Molecular weight.	Concentration.	Molecular weight.
1.1	120	10.6	126
2.6	122	12.7	127
4.0	123	14.3	128
5.1	123	16.6	130
6.6	124	20.5	132
8.1	125	...	...

The deviations from the calculated value, 128, shown by the individual determinations, are not great, but their regularity indicates that these are not due to experimental error, but that a cause inherent in the nature of the experiment, lies at the base of this phenomenon.

A decrease in the molecular weight with rise in temperature is seldom found. Chloral ($CCl_3CHO = 147.5$)

---

[1] E. Beckmann : Ztschr. phys. Chem., **2**, 734 (1898).

in glacial acetic acid, is an example, according to the experiments of Beckmann.[1]

Concentration.	Molecular weight.
0.76	179
2.8	172
5.4	166
10.4	164
14.6	162

Sometimes the values for the molecular weight decrease, at first, with increasing concentration, reach a minimum, and then begin to increase, as is shown by the example given on page 93 of acetic ether in benzene.

The results obtained are very clearly shown in a curve,[2] in which the depressions are plotted as abscissae, and the molecular weights as ordinates.

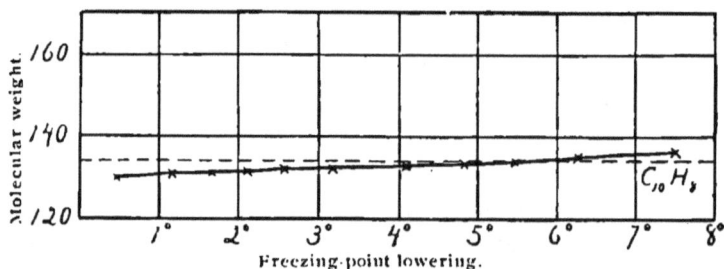

Fig. 19.
Naphthalene in solution in glacial acetic acid.

These curves can be used to derive the molecular weight at infinite dilution, by graphic extrapolation Beckmann[3] found the following values for phenetol ($C_6H_5OC_2H_5 = 122$) in glacial acetic acid.

[1] E. Beckmann : Ztschr. phys. Chem., 2, 724 (1888).
[2] E. Beckmann : Ibid., 2, 719 (1888).
[3] E. Beckmann : Ibid., 2, 732 (1888).

Lowering ················ 0.324° 1.602° 2.522° 4.162° 5.252°
Molecular weight·········   125   136   143   159   171

If the curve is drawn with these values, it will be found to be almost a straight line. The value which would be found for the molecular weight at infinite dilution, can be determined by prolonging the curve to the left, until it intersects the perpendicular drawn through the zero point. The distance cut off on this perpendicular by the curve, corresponds to the molecular weight at infinite dilution. Our example gives the value required, 122.

Other examples are found in the works of Beck-

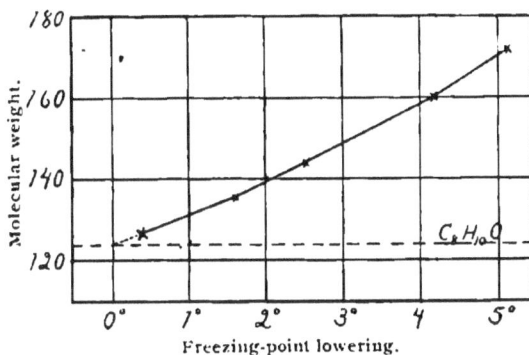

Freezing-point lowering.

Fig. 20.
Phenetol in solution in glacial acetic acid.

mann. The reason for this procedure lies in the fact that the laws of osmotic pressure, and the methods for determining molecular weights based upon them, hold most rigidly in very dilute solutions. The true value of the molecular weight would, therefore, be expected at infinite dilution, which cannot be tested experimentally with the thermometer.

This method of graphically extrapolating the value for the molecular weight, is to be recommended, when the curve is not a straight line nearly parallel to the abscissæ, but when it is a straight line deviating widely from this direction, or is bent.   In every case, several observations must be made in very dilute solutions (depression 0.05° to 0.25°), because many curves are strongly bent in this part of their course, especially those of the acids and oximes, (Figs. 22 and

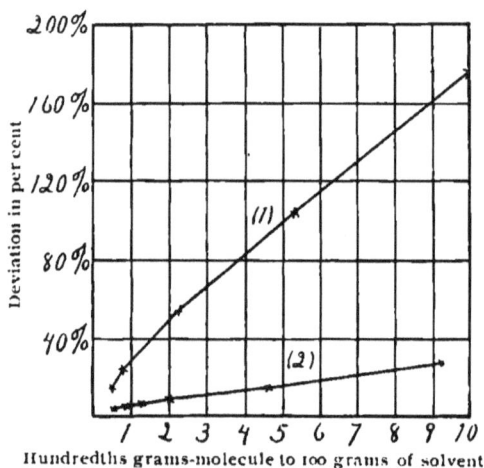

Fig. 21.

(1) *p*-oxybenzaldehyde.   (2) *p*-Cresol in solution in naphthalene.
(Drawn by the method of Auwers.)

23) and the neglect of this portion of the curve would lead, on extrapolation, to erroneous results.

Auwers[1] undertook a recalculation of the numerical values, in order that the results obtained with different substances in the same solvent, could be so brought

[1] K. Auwers: Ztschr. phys. Chem. 21, 339 (1896).

together in a table of curves, that the different curves would be directly comparable. The concentrations, multiplied by 100 and divided by the molecular weight (hundredths gram-molecular weight substance to 100 grams solvent), were taken as abscissæ, and the percentage deviations of the molecular weight found, from the theoretical, as ordinates. The results obtained for p-oxybenzaldehyde and p-cresol, are drawn in the above table of curves, according to the method of Auwers. Many curves thus drawn, are contained in the tables in the work of Auwers already mentioned. In these, any regularities which exist, can be easily recognized, but they suffer from the unavoidable defect, that the initial portions of the different curves lie very close to one another, so that it is not very easy to keep them apart. But as is shown in the above mentioned publication, a larger number of curves can be drawn with the same coordinates, without becoming confused, if the drawing is neatly done.

ELECTROLYTIC DISSOCIATION

The investigation of aqueous solutions of salts, leads to very irregular results, especially those of the strong inorganic acids and bases, and further, solutions of strong acids and bases themselves. Sodium chloride, for example, in dilute solution, gives a depression almost double that which would be expected from the formula, whereas, the molecular weight calculated from this, would be only a little more than half the theoretical. Some salts, such as barium chloride, magnesium chloride, potassium sulphate, show still greater deviation.

This phenomenon has been explained by Arrhenius[1] with the aid of the following assumption. The substances which show this irregularity, on dissolving, dissociate into part molecules, the so-called ions : *c. g.*,

Sodium chloride, NaCl, into Na - Cl,
Potassium sulphate, K$_2$SO$_4$ into K   K   SO$_4$, or
into K - KSO$_4$.

The ions are to be distinguished from the free atoms, in that they are charged with large quantities of electricity. The number of dissolved molecules contained in the solution, is largely increased by this dissociation, since the ions act as molecules. In the case of sodium chloride it was almost doubled, consequently the mean molecular weight is almost half that which would be expected from the formula NaCl ; NaCl =

58.5, mean molecular weight of the ions $\dfrac{\text{Na - Cl}}{2} = 29.2$.

The dissociation is greatest in very dilute solutions ; the greater the concentration of a solution, the smaller the fraction of dissociated molecules. The amount of the dissociation depends, further, upon the chemical properties of the substance. The weak acids and weak bases are not the least dissociated, their salts are more strongly dissociated, while the strong acids and bases and their salts are the most strongly dissociated. Weak acids and weak bases give, therefore, nearly normal values for the molecular weights ; examples are, lactic acid, formic acid, butyric acid, ethylamine. The halogen compounds of bivalent mercury, as well

---

[1] Svante Arrhenius : Ztschr. phys. Chem., 1, 631 (1887).

as the cyanide, are but little dissociated. The corresponding cadmium salts are somewhat more dissociated, and the corresponding zinc salts still more.

Formic acid,[1] like water, is a strong dissociating agent, though it is not as strong as water. Methyl alcohol[2] dissociates less strongly than formic acid.

The amount of dissociation can be ascertained from a determination of the molecular weight by the freezing-point method, in a manner similar to that employed in ascertaining dissociation from the determination of vapor-density. Let M represent the molecular weight calculated from the formula, M' the molecular weight found (smaller than M), n the number of ions into which a molecule dissociates, and γ the degree of dissociation; i. e., the fraction of the molecules originally present in the undecomposed condition, which is broken down into ions, then we have:

$$\gamma = \frac{M - M'}{(n - 1)M}.$$

All solutions which conduct the electric current contain such dissociated molecules. Indeed, it is the ions which make conductivity possible; the larger the number of ions the greater the conductivity of the solution. Upon this fact is based a second method for determining the amount of dissociation, which gives the same values as the cryoscopic method, for molecules which break down into only two ions.

[1] H. Zanninowich-Tessarin : Ztschr. phys. Chem., **19**, 251 (1896).

[2] G. Carrara : *Ibid.*, **21**, 680 (1896).

Since the experimental errors of this electrical method are smaller, it gives more accurate results. The results of the two methods are different for substances which break down into more than two ions.

Since, then, only conductors of the current break down into ions, this kind of dissociation has been termed "Electrolytic dissociation."[1] It must, however, be clearly borne in mind, that this condition of dissociation is not brought about by the passage of the current through the liquid, but is completely independent of it, existing in the liquid through which no current is passing. The capability of a liquid to conduct the current, is but an expression of the fact that it contains free ions.

Another kind of splitting-up which certain substances undergo in solution, is to be distinguished from electrolytic dissociation, although it is very similar to the latter in its effect upon the molecular weight found. It is the breaking down which certain chemical compounds undergo, when formed of two substances loosely held together. Quinhydrone decomposes in solution into quinone and hydroquinone, losing its intense color. Belonging to this same class are the compounds of the hydrocarbons with picric acid.[2] Chloral hydrate[3] belongs also in this class, since in solution in acetic acid, it gives considerably smaller values than would correspond to the formula $CCl_3.CH(OH_2$, evi-

---

[1] Fitzgerald proposed the name ionization : Ztschr. phys. Chem., 7, 400 (1891).

[2] R. Anschütz : Ann. d. Chem., 253, 343 (1889); E. Paterno and R. Nasini : Ber. d. chem. Ges., 22, R. 644 (1889).

[3] E. Beckmann : Ztschr. phys. Chem., 2, 724 (1888).

dently because it is partially broken up into chloral and water. This decomposition is not shown in aqueous solution, since the dissociation is driven back by the large quantity of one of the possible products of dissociation. The same phenomenon is exhibited by chloral alcoholate,[1] which decomposes into its components in benzene, and still more in acetic acid and water. Finally, attention should be called to salts containing water of crystallization, which lose their water of crystallization in aqueous solution. In none of these cases does electrolytic dissociation take place.

A substance like ammonium chloride can decompose by dissociation, in two ways. In the gaseous condition it is split up into ammonia and hydrochloric acid :

$$NH_4Cl \quad NH_3 + HCl.$$

In aqueous solution, on the other hand, the components are the ammonium and the chlorine ions :

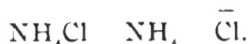

$$NH_4Cl \quad \overset{+}{NH_4} \quad \overset{-}{Cl}.$$

The determination of the molecular weight gives half the true value, in both cases, but as the equations just given show, this is explained differently in the two cases.

More Complex Molecules. — While the electrolytic dissociation in aqueous solutions, can lead to smaller molecular weights than would be expected from the formula of the substance, the reverse can take place in some other solvents ; *i. e., a condensation of several*

---

[1] E. Beckmann : Ztschr. phys. Chem., **2**, 724 (1888).

simple molecules to form a more complex molecule.
The determination of the molecular weight in this
case, gives values which are too large. The term
"association" has recently been proposed for this kind
of condensation. This phenomenon depends upon the
nature of the solvent, the nature of the dissolved substance, and the concentration.

The following solvents favor the formation of more
complex molecules :

ASSOCIATING SOLVENTS.

Anethol,	Methyl oxalate,
Azobenzene,	Naphthalene,
Benzene,	Nitrobenzene,
Bromoform,	Phenanthrene,
Dimethylaniline,	p-Propylanisol,
Diphenyl,	p-Toluidine.
Diphenylmethane,	

On the other hand, we find the simple molecules in
the following solvents :

NON-ASSOCIATING SOLVENTS.

Acetoxime,	Phenol,
Formic acid,	Phenylpropionic acid,
Aniline,	Stearine,
p-Bromphenol,	Stearic acid,
Cetyl alcohol,	Thymol,
Chloral alcoholate,	Urethane,
Acetic acid,	Urethylane.
p-Cresol,	

Acetophenone has a weaker action than these, while
the strongest of the above are formic and nitric acids.

The hydrocarbons are, then, most favorable to the
formation of complex molecules, and solvents containing hydroxyl to simple molecules.

If it is a question of establishing the simplest molecular weight, which is of the most interest to the chemist, a solvent with dissociating .power is chosen for this purpose from the second table. But this precaution is necessary only when the substance whose molecular weight is to be determined, tends to form complex molecules.

Experience has shown that only those substances which contain hydroxyl, and such substances which, through desmotropism, can easily pass over into hydroxyl compounds, form double molecules and still larger molecular aggregations. Very typical of this is the tendency of the organic acids to form double molecules, a tendency which had already been earlier observed in vapor-density determinations. The tendency towards the formation of molecular aggregates was first shown for the remaining classes of substances containing hydroxyl (oximes, alcohols, phenols, acid amines, etc.), through cryoscopic observations. These substances form, therefore, in addition to the ordinary simple molecules, which in general were to be observed in the gaseous state, one or more kinds of more complex molecules, which have been formed through a more or less stable union of simple molecules. This phenomenon is of great significance for our knowledge of the nature of the liquid state of aggregation, or still more, for our knowledge of the condition of dissolved substances.

Finally, the formation of complex molecules is largely dependent upon the concentration. In very dilute solutions, the cryoscopic determinations of the

molecular weights of the classes of substances indicated in the solvents contained in Table I, either give
values which correspond to the simple molecule, or
which are only a little higher.   In the latter case, the
simple molecular weight is always obtained by extrapolating the curve. A greater or less number of
simpler molecules unite to form double molecules, and
still higher molecular aggregates, in somewhat more
concentrated solutions. Higher values are accordingly found for the molecular weights, which increase
with increase in the concentration.

The acids and oximes form double molecules in
concentrated solutions, as already observed. As far

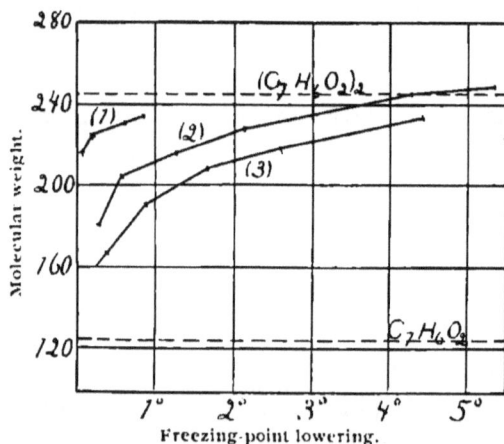

Freezing-point lowering.
Fig. 22.
(1) Benzoic acid in solution in benzene ; (2) In naphthalene ;
(3) In azobenzene.

as investigations have been carried up to the present,
there is no reason to think that they form still more
complex molecules. Therefore, their dissociation

curves represent a simple dissociation of double molecules into single molecules, similar to the curves which represent the dissociation of vaporized substances, as determined by vapor-density methods.

The remaining substances containing hydroxyl, differ from the acids and oximes in that, as far as is known up to the present, they form larger molecular complexes. Their curve is, therefore, not parallel to the axis of the abscissæ at any point, but there is an

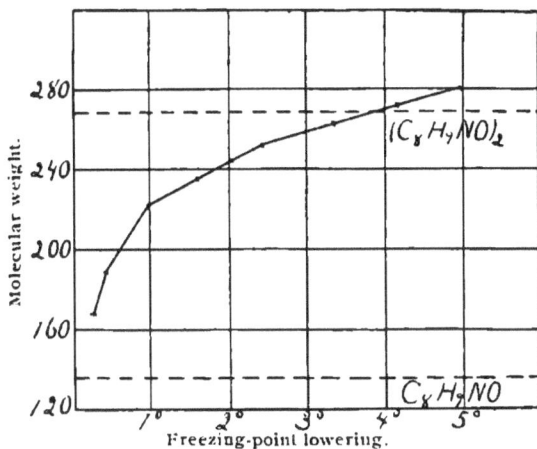

Fig. 23.
Acetophenoneoxime in solution in benzene.[1]

indication that such would be the case at greater concentration. Let a curve of ethyl alcohol,[2] and that of p-oxybenzaldehyde be chosen as examples.

---

[1] From E. Beckmann's determination : Ztschr. phys. Chem., 2, 717 (1888).

[2] It has not been shown in the investigation of ethyl alcohol in solution in benzene, whether a mixture of alcohol and benzene may not perhaps have separated, which would explain the higher molecular values found.

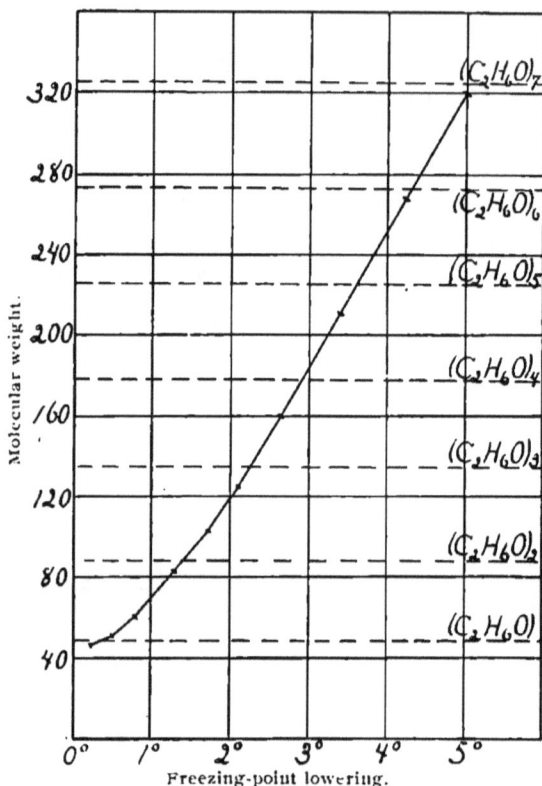

Fig. 24.
Ethyl alcohol in solution in benzene.[1]

An exception is furnished among the alcohols in the derivatives of vinyl alcohol,[2] $CH_2=CHOH$, in which the carbon atom, in combination with the hydroxyl group, is united by double union with a neighboring carbon atom. All substances which we at present regard as being built up according to this

[1] E. Beckmann : Ztschr. phys. Chem., 2, 728 (1888).
[2] K. Auwers : Ibid., 15, 40, 41 (1894).

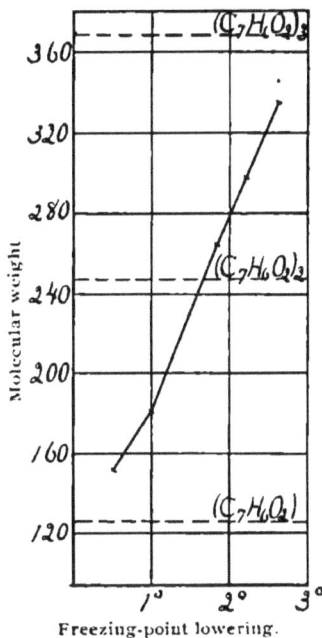

Freezing-point lowering.
Fig. 25.
p-oxybenzaldehyde in solution in naphthalene.[1]

formula, give the same values for the molecular weight at different concentrations. Examples are : Triphenyl-vinyl alcohol, $CH_3COH\ OHCCH_3$

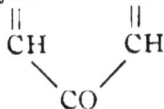

$$\begin{array}{ccc} \| & & \| \\ CH & & CH \\ & \diagdown\ CO\ \diagup \end{array}$$

diacetyl acetone, dibenzoyl acetone, formyl camphor with the group,

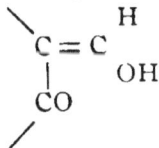

$$\begin{array}{c} \diagdown\quad\quad H \\ C = C \\ | \quad\quad OH \\ CO \\ \diagup \end{array}$$

_____

[1] K. Auwers : Ztschr. phys. Chem., 18, 606 (1895).

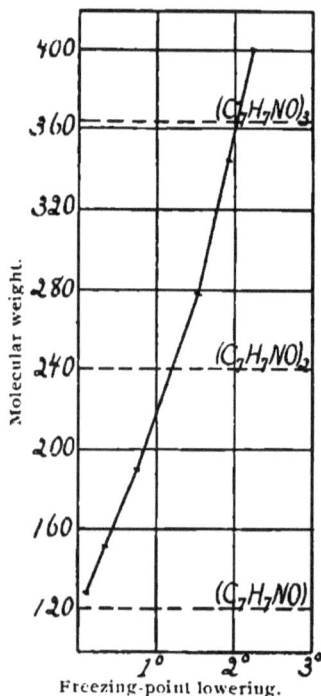

Freezing-point lowering.
Fig. 26.
Formanilid in solution in benzene.[1]

It is interesting to note that the thioalcohols[2] and thiophenols also behave normally.

The acid amides are to be added to the substances which contain hydroxyl.[3] This indicates that, in these cases, a rearrangement has taken place through the migration of the hydrogen ion, which led to the formation of a hydroxyl group; c. g., diphen-

[1] E. Beckmann : Ztschr. phys. Chem., 12, 711 (1893).

[2] K. Auwers: Ibid., 12, 693; 711 (1893).

[3] K. Auwers : Ibid., 15, 45, 50 (1894); A. Lachmann : Ibid., 22, 170 (1897).

oxyacetamide, $(C_6H_5O)_2CH.CO.NH_2$, passes over into $(C_6H_5O)_2CH.C.NH.$
OH

This phenomenon has been investigated most advantageously with the acid derivatives of the organic bases, especially with the formic acid derivatives, of which a large number has been studied by Auwers,[1] e. g., formanilid, $C_6H_5NH.COH$, behaves as if it were
$C_6H_5N : CH$
OH

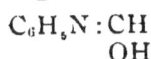

To these are to be added some of the amides of carbonic acid, such as ordinary urethane,

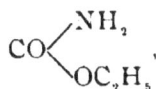

$$CO{<}\begin{matrix}NH_2\\OC_2H_5\end{matrix}\quad,$$

which acts like a substance,

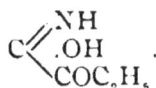

$$C{<}\begin{matrix}NH\\.OH\\COC_2H_5\end{matrix}\quad.$$

The fact that acid derivatives of the secondary bases behave normally,[2] indicates that the cause of the anomaly is a rearrangement, with the formation of a hydroxyl group; e. g.,

$$\begin{matrix}C_6H_5.N.C_6H_5\\CO\\H\end{matrix}$$

formyldiphenylamine and

$$\begin{matrix}C_6H_5.N.CH\\CO\\H\end{matrix}$$

formylmethylaniline.

[1] K. Auwers: Ztschr. phys. Chem., 15, 49 (1894).

[2] K. Auwers: Ibid., 15, 44; 45 (1894).

These substances contain in combination with the nitrogen, no available hydrogen which can take part in the formation of a hydroxyl group. That it is not the imide group which favors the molecular condensation, is shown by the fact that pure imide substances, which cannot pass over into hydroxyl compounds by the migration of a hydrogen atom, behave normally; *e. g.*, monomethylaniline, monoethylaniline,[1] acridine,[2] skatol.[2]

Finally, the nitroso compounds of the tertiary aromatic bases,[3] show inclination to form molecular complexes, although, according to our present view, they do not contain a hydroxyl group. It should, however, be observed, that the constitution of this remarkable class of substances is not completely cleared up.

On the other hand, nitric acid[4] dissolved in nitrobenzene gives normal values, nothwithstanding that it contains a hydroxyl group.

The class of the phenols has been studied with unusual thoroughness and with success, in the beautiful work of Auwers[5] and his pupils, using naphthalene as a solvent. It has been shown that the capability of the phenols to form complex molecules, depends essentially upon the substituents already present in the molecule, these either favoring or hindering condensation, depending upon their position and their chem-

---

[1] H. Hof : Dissertation, Erlangen 1895, p. 26.
[2] K. Auwers : Ztschr. phys. Chem., **12**, 712, 713 (1893).
[3] K. Auwers: *Ibid.*, **12**, 715 (1893).
[4] H. Hof : Dissertation, Erlangen 1895, p. 11.
[5] K. Auwers : Zeit. phys. Chem., **18**, 595 (1895) ; **21**, 337 (1896).

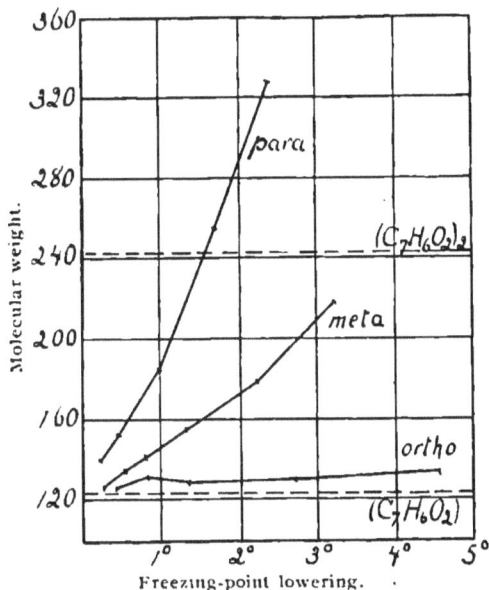

Freezing-point lowering.
Fig. 27.
The three oxybenzaldehydes in solution in naphthalene.[1]

ical composition. Ordinary phenol is to be classed with the alcohols, as also the parasubstituted phenols. The meta-derivatives show a somewhat lighter tendency in this direction. It has completely disappeared in the ortho-derivatives. These behave like substances which do not contain hydroxyl, and give constant values with increasing concentration.

In addition to the position, the composition of the substituents is to be taken into account. The aldehyde group, — COH in the para position to the hydroxyl group, is most favorable to the formation of complex molecules. Next comes the carboxethyl

[1] K. Auwers: Ztschr. phys. Chem., 18, 605, 606 (1895).

group, $-$COOR, then the cyanogen group, $-$CN, and then the nitro group, NO_2. Then follow the halogen atoms in the order, I, Br, Cl, and the least active of all are the alkyl groups. On the contrary, values are obtained which deviate but slightly from the calculated molecular weight, if one of these groups is in the

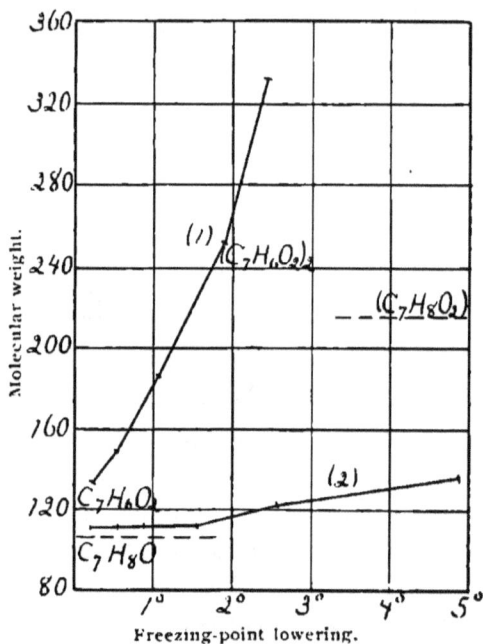

Freezing-point lowering.

Fig 28.

(1) *p*-Oxybenzaldehyde, and (2) *p*-cresol in solution in naphthalene.[1]

ortho position to the hydroxyl group. This deviation is least with the aldehydes and nitro compounds. Larger deviations appear, corresponding to the order of succession, on introducing the remaining substituents. In conformity with the rule that the para-

[1] K. Auwers : Ztschr. phys. Chem., **18**, 599, 606 (1895).

phenols tend to form double molecules, and the ortho-phenols single phenols, it can be said that the aldehyde and nitro groups tend to give normal results. The fairly normal curve of *p*-cresol, and the rapidly

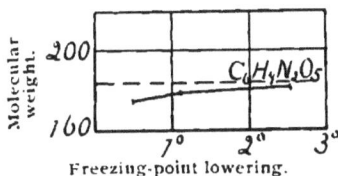

Fig. 29.

o-p-dinitrophenol in solution in naphthalene.[1]

ascending curve of *p*-oxybenzaldehyde, may be taken as examples.

If there are several substituents present in the phenol, the one in the ortho position to the hydroxyl

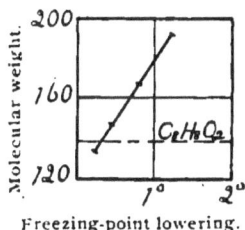

Fig. 30.

p-oxy-*m*-methylbenzaldehyde in solution in naphthalene.[2]

has the strongest action. Its effect can be overcome only when it itself is a weakly acting substituent (alkyl), and a strongly acting substituent stands in the para position. The curve of *o-p*-dinitrophenol is, therefore, fairly normal; that of *p*-oxy-*m*-methyl benzaldehyde rises rapidly.

[1] K. Auwers: Ztschr. phys. Chem., **18**, 603 (1895).

[2] K. Auwers: *Ibid.*, **18**, 606 (1895).

These regularities were first found for solutions in naphthalene. It is probable that they hold, in the same way, for solutions in the other non-dissociating solvents. At least, the few results found for the phenols in benzene do not oppose this.

Analogous regularities may, perhaps, be shown to exist for still other classes of substances, so that the course of the cryoscopic curve may be occasionally used to advantage in determining constitution. But our knowledge of the effect of constitution on the tendency towards polymerization is, at present, for other classes of substances, too slight to permit any use being made of it for determining constitution.

Auwers[1] has, in a brilliant manner, made use of the regularities found by him in the investigation of the phenols, for determining the constitution of the oxyazo compounds. It was shown that the paraoxyazo compounds are to be regarded as containing hydroxyl; e. g., benzeneazophenol, as,

$$C_6H_5N = NC_6H_4OH.$$
$$1 : 4$$

On the other hand, the orthooxyazo compounds are to be regarded as the phenylhydrazones of the orthoquinones; e. g.,

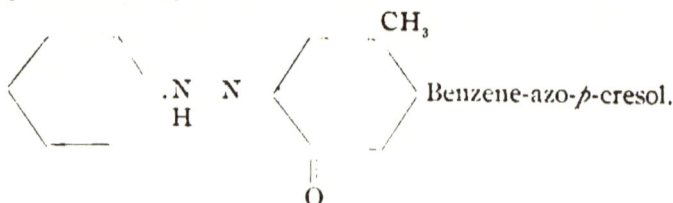

Benzene-azo-p-cresol.

[1] K. Auwers and K. Orton : Ztschr. phys. Chem., 21, 355 (1896).

It is a general rule, that in those classes of substances which give abnormal molecular weights, the deviations are greatest in the lowest members of the series, and least in the highest members. Auwers[1] found this rule for alcohols, Hof[2] for anilides. This same relation was shown in the abnormal vapor-densities, as already observed.

To recapitulate briefly, it was found that the concentration curves for acids and oximes, which were

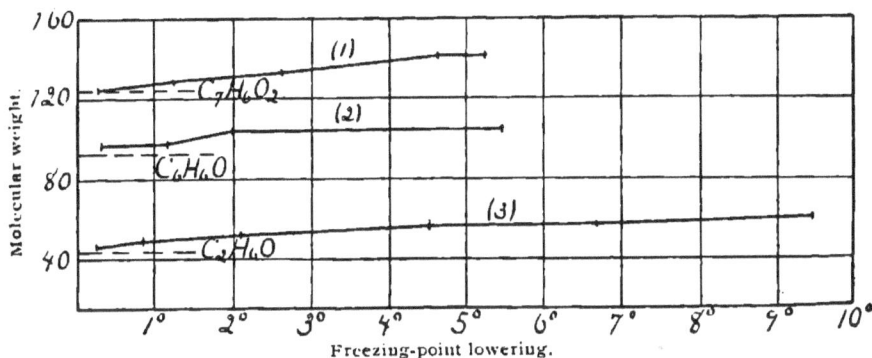

Freezing-point lowering.
Fig. 31.
(1) Benzoic acid; (2) phenol; (3) ethyl alcohol in solution in glacial acetic acid [3]

investigated in solvents of the first table, were bent, their concave side being turned towards the abscissæ. For all other substances it is represented by a straight line, which, as a rule, is only a little inclined. But, on the other hand, for substances containing hydroxyl it rises more or less rapidly, depending on the concentration, and only in a few cases, behaves normally.

[1] K. Auwers : Ztschr. phys. Chem., 12, 705 (1893).
[2] H. Hof : Dissertation, Erlangen 1895, p. 19.
[3] E. Beckmann : Ztschr. phys. Chem., 2, 732 (1888).

The simple molecular weight for every one of these "*abnormal substances*," *can be derived from a series of molecular weight determinations plotted in a curve.* But the interpretation of one singular molecular weight determination is often difficult, and at times impossible.

Curves which are always normal, and therefore almost horizontal, are found in solvents of the second table. Benzophenone and benzoic acid anhydride, in

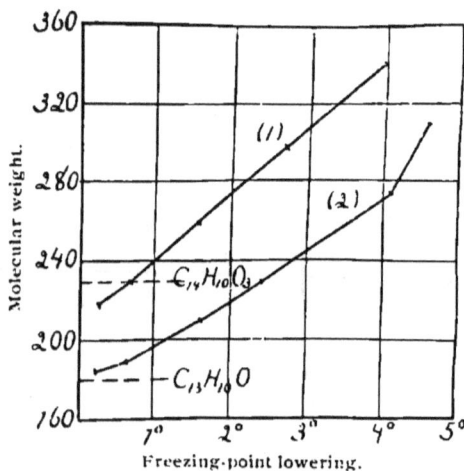

Freezing-point lowering.

Fig. 32.
(1) Benzoic acid anhydride; (2) benzophenoue, in solution in glacial acetic acid.[1]

solution in glacial acetic acid, are exceptions. Their curves rise rapidly, according to the investigation of Beckmann,[1] and look like a curve of an alcohol in benzene. Dicyandiethyl, $C_6H_{10}N_2$, in benzene, behaves in the same way. No explanation, whatever, of these remarkable exceptions, has thus far been furnished.

_____

[1] E. Beckmann : Ztschr. phys. Chem., **2**, 722 ; 732 (1888).

DETERMINATION OF MOLECULAR WEIGHTS BY THE BOILING-POINT METHOD

Three fundamental statements can be made at first in considering the boiling-point method :

(1) A solution boils higher than the solvent.

(2) The rise in boiling-point is proportional to the concentration.

(3) Equimolecular solutions in the same solvent, show the same rise in boiling-point.

The generalizations which have led to the boiling-point method, have not been found directly in this form, but were derived, at first, rather from observations on the vapor-pressure of solutions and solvents, by Wüllner, Ostwald, and Raoult. But diminution in vapor-pressure and rise in boiling-point are proportional, within limits which are not too wide. This historical development can thus be explained : There are disturbing complications involved in the determination of the boiling-points of a solution, which, only very recently, have been overcome, while there are no difficulties in the method involved for determining vapor-pressure. The thermometer must dip into the solution to determine its boiling-point. But overheating, warmer and colder currents, small changes in the amount of heat supplied, produce considerable variations in the position of the thermometer, which can amount to more than a degree.

It was Beckmann,[1] who overcame these difficulties,

---

[1] E. Beckmann : Ztschr. phys. Chem., 4, 539 (1889).

after Raoult, some years before, had carried out experiments along this line, which, at first, were only
partially successful. Beckmann accomplished his aim
through two new improvements. First, he fused a
platinum rod into the bottom of the boiling-vessel,
which prevents bumping, and second, he partly filled
the boiling-vessel with some coarsely granular material,
which prevents overheating. These two devices work
so well that it was no longer difficult to keep the
boiling-point of a solution constant for several hours,
to within a few thousandths of a degree. The best
guarantee that Beckmann has completely accomplished
his aim, and that an overheating of the boiling liquid
is entirely avoided by his method, is that a pure substance shows the same boiling-point, whether the thermometer is heated by the vapor of the boiling liquid,
or whether it dips into the liquid itself, provided the
Beckmann apparatus is used.

The "*Molecular rise in boiling-point*," or "*Boiling-
point constant*" K', is used as the value for comparison in calculating molecular weights. K' is the
number of degrees, which the boiling-point of 100
grams of the solvent is raised by a gram-molecular
weight of the substance dissolved in it. An equation
for the determination of the molecular rise in boiling-
point, is derived from the above fundamental propositions, exactly as in the freezing-point method.

An experiment has shown that the boiling-point of
L' grams of solvent, is raised E degrees by dissolving
S' grams of the substance in it. By dissolving one
gram of the substance, the rise $\dfrac{E}{S'}$ would be obtained

(fundamental proposition 2). If one gram of the sol·
vent had been employed, the rise would have been
$\dfrac{EL'}{S'}$; and if 100 grams of the solvent had been em-

ployed, $\dfrac{EL'}{100\,S'}$. Finally, if a gram-molecular weight
of the substance had been dissolved, the rise would
have been $\dfrac{EL'M}{100\,S'}$. From the third fundamental proposition this expression, for one and the same solvent
in which any substance is dissolved is :

$$K' \quad \dfrac{EL'M}{100\,S'}.$$

If the molecular weight of the dissolved substance
is known, the constant K' is found from a determination of the rise in boiling-point which the substance
produces.

Example: 0.5475 gram benzil, ($C_{14}H_{10}O_2 = 210$) in
38.09 grams of benzene, gave a rise in boiling-point of
0.174°.

$$K' \quad \dfrac{0.174}{100} \cdot \dfrac{38.09 \vee 210}{0.5475} \quad 25.4.$$

The mean of a large number of experiments is
26.1. Furthermore, the boiling-point constant K'[1]

---

[1] The constant K' can, further, be approximately derived from
the law of Trouton. $K' = 0.00096\,TM$; T is the absolute boiling
temperature, and M the molecular weight of the solvent. E. Beckmann, G. Fuchs, and V. Gernhardt : Ztschr. phys. Chem., **18**, 473
(1895). In the same place a method is described for deriving the
heat of vaporization, from a determination of the boiling-point of a
solvent under different pressures. The boiling-point constant is
then calculated from this.

can be derived in the same manner as the freezing-point constant K. Let T_s be the absolute temperature, W_t the latent heat of vaporization, then :

$$K' \quad \frac{0.0198 \; T_s^2}{W_t}.$$

On the contrary, if the heat of vaporization of a substance is not known, this is derived from the determination of the rise in boiling-point which is produced, by dissolving in it another substance of known molecular weight. K' is at first derived from this observation, and finally, W_t found by introducing this into the last formula. This method has been used by Beckmann and Fuchs[2] for determining the heat of vaporization of a large number of substances.

If the boiling-point constant of a solvent is known, it can be used for determining the unknown molecular weights of other substances which are soluble in this solvent. The molecular weight is obtained from the equation :

$$M \quad \frac{100 \; S'K'}{EL'}.$$

This method for determining molecular weights was discovered and developed by Raoult.[3] It has been used by Walker, Löb, Tammann, Will, and Bredig, in a modified form adapted to the chemical laboratory. Raoult, already in 1878, had demonstrated the possibility of obtaining the same results through boiling-

---

[1] Sv. Arrhenius: Ztschr. phys. Chem., **4**, 550 (1889).

[2] E. Beckmann and G. Fuchs: *Ibid.*, **18**, 473 (1895).

[3] F. M. Raoult: Compt. rend., **87**, 167 (1878); **103**, 1125 (1886); **104**, 976, 1430 (1887); **105**, 857 (1887); **107**, 442 (1888); Ann. Chim. Phys. [6], **15**, 375 (1888); Ztschr. phys. Chem., **2**, 353 (1888).

point determinations, as through observations of the vapor-pressure, but regarded the latter method as preferable even up to 1889. In the same year, Wiley employed the boiling-point method for determining the molecular weights of some salts; but Beckmann[1] was the first to develop it into a really practicable method. He submitted the apparatus constructed by himself, to the German Society of Scientists, on September 21, 1889, and shortly afterwards published a description of it. At the same time, Arrhenius[2] furnished the theoretical basis for the method. An improved form of apparatus was described by Beckmann[3] in 1891.

We owe the principle of both the freezing-point and boiling-point methods to the physicist, Raoult, the practical development to the chemist, Beckmann.

The Simple Boiling-point Apparatus of Beckmann.[4] — The boiling liquid is contained in a test-tube (a),

[1] E. Beckmann : Ztschr. phys.Chem., **3**, 603(1889); **4**, 532 (1889).

[2] Sv. Arrhenius : *Ibid.*, **4**, 550 (1889).

[3] E. Beckmann : *Ibid.*, **8**, 223 (1891).

[4] The description of the first Beckmann boiling apparatus, without vapor-jacket, which is adapted only for low-boiling solvents, is given in Ztschr. phys. Chem., **4**, 539 (1889). An account of experiments performed with it is given by E. Beckmann in Ztschr. phys. Chem., **6**, 437 (1890). This apparatus has been recently improved and also adapted to high-boiling solvents ; E. Beckmann : Ztschr. phys. Chem., **21**, 245 (1896). The apparatus described in the text, with vapor-jacket, was described by E. Beckmann in 1891 in Ztschr. phys. Chem., **8**, 223 (1891). Some modifications, together with a criticism of changes proposed in other places, in Ztschr. phys. Chem., **15**, 656 (1894). For carrying out boiling-point experiments under changing pressure, the apparatus of E. Beckmann and G. Fuchs has been somewhat modified. Ztschr. phys. Chem., **18**, 492 (1895) : E. Beckmann : Ztschr. phys. Chem., **18**, 661 (1894).

the "boiling-vessel") provided with a side tube. It has the same form, and also nearly the same dimensions as the freezing-vessel in the freezing-point apparatus. A thick platinum rod about ½ cm. long, is fused into the bottom of the boiling-vessel with the aid of fusion glass, to prevent the boiling liquid from bumping. The rod projects externally only a little beyond the surface of the glass, and is notched along that portion which projects into the vessel. The formation of bubbles takes place easily on the sharp edges, because small bubbles of vapor, which really produce the boiling, adhere to these. Moreover, heat is conducted from the outside into the interior of the liquid, through the platinum rod, which can be replaced also by a rod of red fusion glass. This boiling device can be dispensed with without any serious disadvantage.[1] A Beckmann thermometer is placed in position, by means of a cork, which had been extracted with ether before using, so that the mercury bulb stands about 3½ to 4½ cm. from the bottom of the boiling cylinder. A light glass condenser with a straight condenser tube, is fitted to the side tube with a cork which has also been cleansed with ether. The boiling-vessel fits loosely into the hollow of the glass boiling-jacket (*b*), so that it scarcely protrudes below. The inner space of the boiling-jacket, closed on all sides, is connected above (left in the figure) with a small condenser which is bent upward. The effect of the temperature of the air is excluded by introducing

[1] E. Beckmann : Ztschr. phys. Chem., **15**, 661 (1894).

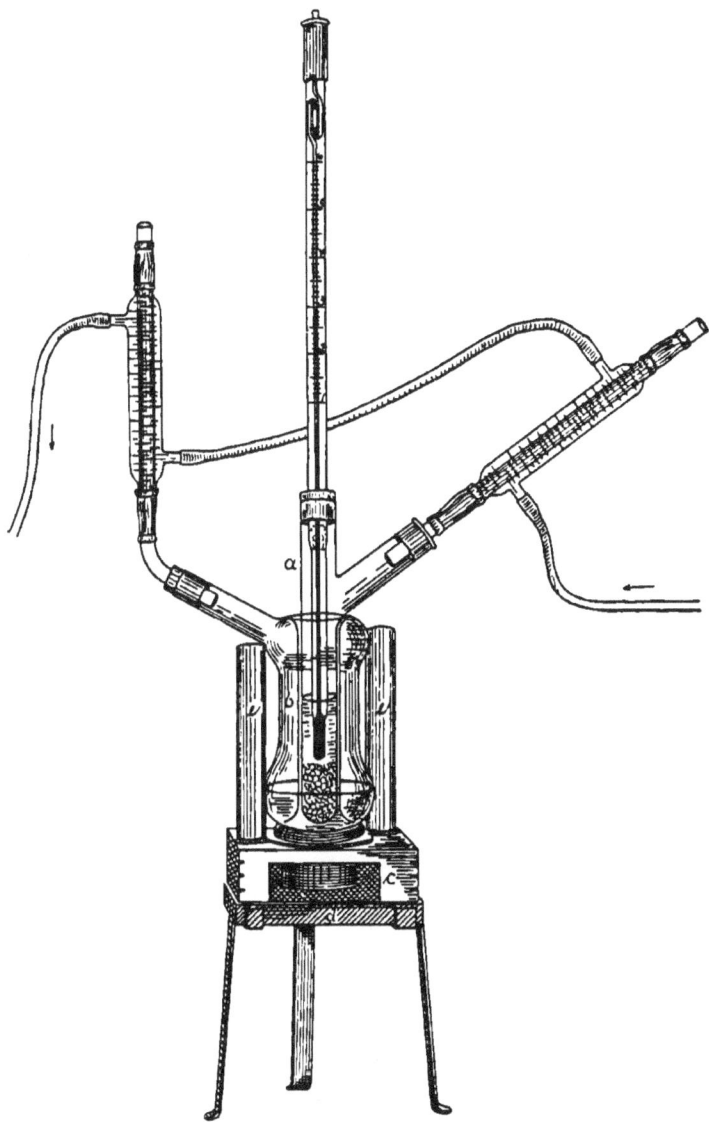

Fig. 33.
Beckmann boiling-point apparatus. One-fifth natural size.

into the outer vessel the same solvent as in the inner.
This device is of great advantage for experiments
with high-boiling solvents, boiling from 100° to 200°.

Both systems were placed on a heating-box, made
with strong asbestos board held together with wire
clamps and water-glass. Its arrangement is shown in
Figs. 33 and 34. The bottom rests on an iron support (Fig. 33, *d*), together with the side walls 5 cm.

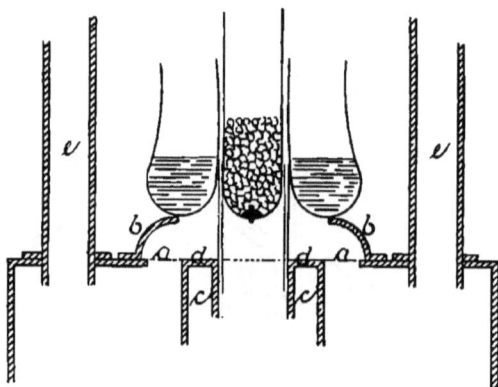

Fig. 34.
Diagonal section through the heating-box. One-third natural size.

in height, a mica window[1] having been inserted into
one of these. It has an opening in the center 8 cm.
in diameter, which is covered with a piece of brass
gauze (Fig. 34, *a*). On the shelf on which the brass
gauze rests, is placed a ring of asbestos board, bent
upward, as shown in the figure (Fig. 34, *b*), protruding

---

[1] The object of the mica window is to allow the flames to be
observed, and is strongly to be recommended when the circular
burner, to be described later, is used. When a Bunsen burner is
used, the flame can be observed just as well from below, and a box
without a mica window is preferable, since it is more durable.

over the gauze, and having a hole in the center about
6 cm. in diameter. The boiling-jacket with boiling-
vessel is placed upon this. Two concentric rings of
asbestos board, 3 and 5 cm. in diameter (Fig. 34, *c*),
and 3 cm. in height, and fastened below, are placed
near the center of the gauze. They are united at the
level of the gauze, by a ring of asbestos (*d*) with a hole
cut in the center. The brass gauze is cut out in the
middle, leaving an opening of the diameter of the
inner ring. This opening is just beneath the bottom
of the boiling-vessel. The rings prevent the direct
action of the surrounding flames on the boiling-vessel
itself. Finally, two asbestos chimneys (*e*) are placed
in two corners of the cover of the heating-box.
These serve to carry off the gaseous products of combustion. Two Bunsen burners, or gas burners of a
form to be described later, are placed below the heating box, so that their flames strike the gauze at places
removed as far as possible from the chimney.

**Carrying Out a Simple Molecular Weight Determi-
nation with the Simple Boiling-point Apparatus of
Beckmann.** — It is recommended in learning the
method to carry out a molecular determination with
benzene as a solvent, and a high-boiling, solid hydrocarbon, perhaps phenanthrene, as the substance to be
investigated.

The thermometer is at first so adjusted that the
top of the mercury column comes to rest between the
divisions o and 1 (say about half way between them),
when the bulb of the thermometer is placed in the
vapor of some benzene, boiling in a wide test-tube.

Sometimes the bulb of the thermometer available, is just far enough from the lower end of the scale, that when the thermometer is properly introduced into the apparatus, this portion of the scale is covered by the cork. In this case the thermometer is, of course, so adjusted, that the top of the column comes to rest just above the cork, when the benzene boils. The thermometer with short bulb, is best adapted to boiling-point measurements. The older freezing-point thermometers, with long bulb (4.5 to 5 cm.), can also be used when necessary. They can, however, be used only with solvents which do not boil above 100°.

Sixteen to seventeen grams of pure, dry, benzene, are weighed off accurately to centigrams in the boiling vessel,[1] then the thermometer with cork put in place, and the position of the thermometer so regulated that its lower end is about 4 cm. from the bottom, and dips well into the benzene. Then some heavy glass beads of about 3 mm. diameter, or garnets of from 2 to 3 mm. diameter, are introduced through ths side tube, until the vessel is filled with them up to the bottom of the thermometer. The thermometer itself must not touch them, since if it did, an exact determination of the zero point would be difficult. The benzene is thus made to rise over the bulb of the thermometer. The properly selected beads or garnets are cleansed, by warming with concentrated hydro-

---

[1] The boiling-vessel must be thoroughly cleansed, otherwise the liquid condensing in the upper part flows down with difficulty, thereby producing an increase in the concentration of the boiling solution. This fault is remedied, by cleaning the vessel with warm concentrated sulphuric acid and potassium chromate.

chloric acid, washing with water and alcohol, and then drying. Platinum is especially well adapted for use, and has been recommended by Orndorff and Cameron,[1] and also recently by Beckmann.[2] The good conductivity of the metal facilitates the equalization of temperature, and makes possible a better adjustment of the thermometer, than glass beads or garnets. Platinum foil is bent for this purpose into balls or tetrahedra, cleansed, and heated to redness before using.

The introduction of a solid filling-material into the boiling-vessel is of the greatest significance, because it prevents a superheating of the boiling liquid. The bubbles of vapor, rising from the bottom, have their movements frequently checked by striking against the glass beads which block up their paths. The bubbles must then pass around these, and in doing so they give up their excess of heat to the liquid, so that when they emerge from the filling-material, they have exactly the temperature of the boiling-point. This exceptionally important means of preventing superheating, without which it would be almost impossible to determine the boiling-point of a solution, has been introduced into practical work by Beckmann.

When the condenser is attached, the boiling-vessel is ready for the experiment. Ordinary benzene is poured into the boiling-jacket, until it is filled half-way up the filling-material in the boiling-vessel, and some

---

[1] W. R. Orndorff and F. K. Cameron : Ztschr. phys. Chem., 17, 638 (1895).

[2] E. Beckmann : *Ibid.*, 21, 248 (1896).

pumice-stone or some pieces of porous porcelain thrown
in.   The proper condenser is attached, and now the
whole apparatus is set up as shown in Fig. 33, the
boiling-vessel being clamped at the upper end in a
retort holder.

If the apparatus thus described, is used at once for
experiment, a considerable fault would be met with.
Some of the hot gases from the flame would ascend
through the brass gauze of the heating box, and rise
well up between the boiling-vessel and boiling-jacket.
Since this would not take place uniformly, the boiling
apparatus would be supplied, now with more, now with
less heat, so that an accurate adjustment of the mercury meniscus would not be possible.   This current of
air is prevented by two devices.   First, a tube made
by rolling up asbestos paper into several layers, is
placed in the inner asbestos ring of the heating box,
as shown in Fig. 34, so that it reaches up into the
boiling-jacket.   The boiling-vessel is now introduced
into the jacket, while the asbestos tube is held in
position below by the finger, that it will not be pushed
down.   The boiling vessel then slips right down into
the asbestos tube.   Second, some loose asbestos is
packed in above, between the boiling-vessel and the
boiling-jacket, so that here, again, is another obstruc-
tion which prevents the gases from rising.   Instead of
using loose asbestos, the boiling-vessel, before it is inserted into the jacket, can be wound around with
several layers of strips of asbestos paper, even up to
the side tube.   This, then, accomplishes the same purpose as the loose asbestos.   Neither of these asbestos

packings is shown in the drawing of the apparatus as a whole, that the figure should not thereby be rendered indistinct.

If the whole apparatus is set up on a wooden table, it is necessary to place a slab of slate beneath it, because the wood would be heated too strongly by the heat radiated down upon it from the heating box.

The apparatus is heated by two Bunsen burners, in the manner already described. Two sheets of metal, bent in the form of semicircles, and which together form a ring about 20 cm. high, are placed around the apparatus to prevent any possible air-currents from affecting the apparatus. If a boiling apparatus with high heating box is used, the burners are placed on small blocks of wood, or the Beckmann adjustable burner, to be described later, is used.

After the apparatus has been set up and filled, which requires from ten to fifteen minutes if everything is at hand and no cleansing is needed, the water is turned into the condenser and the gas lighted. A small flame is used at first. The height of the flame can be somewhat increased after five minutes, so that it reaches the gauze. A cracking of the glass is not to be feared, even if less care is taken. The benzene in the jacket quickly begins to boil. The first bubbles of vapor emerge somewhat later from the filling-material in the boiling-vessel. The rate of boiling is now so regulated, by adjusting the height of the flame, that a drop falls from the condenser of the boiling-vessel every five to ten seconds. It is self-evident that a much more vigorous boiling goes on in the jacket.

It is important to maintain this rate of dropping,
because the thermometer will then quickly reach the
true boiling-point, and show only small fluctuations of
a few thousandths of a degree.   It is well, after the
proper rate of dropping has been established, to read
the thermometer from time to time, say every five
minutes, and to note the readings together with the
time.   An estimate of the rate of movement of the
mercury can best be obtained in this way.   The following fact is to be taken into account in making the
readings: The movement of the mercury in the
capillary is hindered, because of the small diameter
of the capillary, and therefore the mercury does not
accurately respond to small changes in temperature.
The error thus introduced, and to which attention has
been called in discussing the freezing-point method, is
avoided, if just before and during the reading the
upper end of the thermometer is gently tapped with
an empty thermometer case, held in the right hand,
the lens being held in the left.   The capillary is thus
thrown into gentle vibrations.   The tapping is continued from one-half to one minute, and then the reading is made.   A mechanical tapping-device can be
used instead of the hand.   Loomis[1] recommends a
device like the hammer of an electric clock, which
strikes on the top of the thermometer.   The large

---

[1] E. H. Loomis: Wied. Ann., N. F., 51, 506 (1894); W. R.
Orndorff and F. K. Cameron: Ztschr. phys. Chem., 17, 640
(1895); M. Kaehler and Martini: W. Berlin, Nachtragscatalog,
1897; electro- and physicochemical apparatus, p. 37, Nr. 6305;
The mechanic, H. Wittig, in Greifswald, furnishes suitable electrical clock-works for about 7 marks each.

number of slight taps jar the thermometer uniformly throughout, but each blow only so slightly that no oscillation of the capillary can be observed. Consequently, the reading can be made very accurately. The use of this device is strongly to be recommended. Blows on the side can be given by a cork rotating on an axis. One side of the cork must be partly cut off, and the remainder allowed to strike the thermometer once in every turn. The cork is driven by a cord attached to a small water turbine. Turbine and cork are fastened to a wooden lath about 40 cm. long, supported on a stand.

The electrical clock-work already mentioned, is preferable. But it is sufficient in an ordinary molecular weight determination, to tap with the hand in the manner described.

It usually requires one hour, but often two, and sometimes, indeed, several hours, to obtain a constant temperature in the boiling-vessel. If one is already familiar with the method, it is sufficient to observe the position of the mercury thread every quarter of an hour, and to begin the readings proper, only when the mercury has about come to rest.

The tablets, molded in exactly the same manner as described in the chapter on the freezing-point method, are meanwhile gotten ready and weighed.

When the thermometer ceases to rise, and oscillates only slightly about a fixed point on the scale, which is explained by irregularities in boiling, the point is determined accurately by taking the mean of several readings. It should not oscillate more than two- or three-thousandths of a degree, in ten to fifteen minutes, otherwise the experiment cannot be begun. Since the constants of the solvents are, indeed, considerably smaller in the boiling-point method than in the freezing-point, the rise as read on the thermometer in the former case, is less than the fall in the latter. Consequently, the elevations of the boiling-point must be determined with the greatest care, in order to make accurate molecular weight determinations. This is to be taken into account, especially for substances with high molecular weights, and in dilute solutions where a rise of a few hundredths of a degree must be read.

After uniform boiling has been secured in the apparatus, and the boiling-point established, the determinations proper present no further difficulties. The first tablet is slipped into the boiling-vessel, through the tube of the condenser. The temperature falls at first, because the cold tablet, and further the process of passing into solution, absorb heat. But it soon begins to rise above the original position as the substance dissolves, and within five minutes registers the boiling-point of the solution. One should wait a few minutes to determine whether the temperature remains constant, and then make the final reading. A determination usually requires from ten to fifteen minutes, from the time the substance is introduced until the final reading is made. A second tablet is then introduced, and the boiling-point of the new solution determined, the total amount of substance contained in the solution, and the total rise in temperature being taken into account. Four or more deter-

minations can be conveniently carried out in this
manner, according to the purpose of the investigation,
and the solubility of the substance to be investigated.

The boiling of the solvent must not be discontinued
during the introduction of the tablet. Every liquid
which is kept exposed to the air, contains some air in
solution, and the formation of the first vapor-bubbles
during the warming of the liquid, is due to this small
quantity of dissolved air. The air-bubbles adhere to
the walls and to the bottom, but they soon pass over
into bubbles, the air being gradually driven out by the
particles of vapor formed, and replaced by vapor.
Uniform boiling depends upon the presence of such
bubbles.

When the liquid ceases to boil, the small bubbles of
vapor disappear, and if the temperature again rises,
there are no places present from which boiling can
begin. Overheating is the result, and consequent
bumping, which interferes with the experiment.

It is very convenient in introducing the round
tablets, to allow them to roll down through the condenser into the apparatus. In this way, we can be
certain that the tablet enters the boiling-vessel as it is
weighed, without any loss. This is accomplished by
means of the tablet thrower, shown in
Fig. 35, which is made by bending a
piece of sheet-zinc, or brass, into the proper
form. The tablet is placed in the box-
shaped groove open in front, the thrower
seized by the bent portion, lifted up to the
edge of the condenser, and the tablet started rolling, by

Fig. 35.
Tablet thrower.
One-half natural
size.

tilting downward the front, open end of the groove, the
walls of the groove giving it an upright position.   If
the tablets are more than 4 mm. thick, they can no longer
be rolled through the condenser. They are then allowed
to slide down, the axis of the cylindrical tablet in the
axis of the condenser; this being assisted by a long
glass rod. It is desirable that the tablet should be
introduced without allowing the glass rod to come in
contact with the liquid condensed in the lower end of
the condenser. If this should occur, the rod is slowly
drawn backward, and repeatedly touched against the
walls of the condenser, so that the particles of solvent
clinging to it, are wiped off as much as possible, and
thus only a small loss of solvent results. It is simplest,
however, to prepare the tablets of such size that they
can be rolled into the apparatus. When larger
amounts of substance are to be introduced, instead of
one tablet, several are to be thrown in at once. This
has the further advantage that the tablets glide past the
thermometer into the boiling solvent, while larger
tablets remain suspended at the junction of the side
tube, between the thermometer and the wall of the
vessel, and are here gradually dissolved by the condensed liquid as it runs down, and by the rising
vapors, until it can slip down through the intervening
space. It is convenient in the boiling-point method,
that any small particles of the substance which remain
suspended, are washed down by the recondensed solvent. Further, a rapid and uniform mixing of the
solution is effected with certainty by the boiling process, through the movement produced by the bubbles
of vapor as they ascend in the boiling-vessel.

The solution constantly undergoes small changes in concentration, through the drops which fall from the condenser, every such drop increasing in dilution, while the concentration is again increased in the following moment, by the distillation of the solvent high up into the apparatus, until a new drop dilutes it again. These changes in concentration are very perceptible in stronger solutions, and prevent the mercury column from adjusting itself as sharply with more dilute solutions, or with the pure solvent. But since the total elevations observed in these cases are large, an error of reading which does not amount to more than 0.01°, is not of much consequence.

The calculation is made with the aid of the formula already developed, taking into account the following correction. A part of the solvent is always present in the form of vapor, in the upper part of the boiling-vessel, due to the boiling process. Another portion adheres to the walls in small drops, while still another portion is absorbed by the corks with which the vapors come in contact. Consequently, there is always somewhat less solvent below in the boiling-vessel, than had been weighed in it, and the solution is somewhat more concentrated. The error thus produced, is corrected by subtracting a small amount, about 0.15 to 0.2 gram, from the weighed mass of the solvent. Only when water is used should about 0.35 gram be subtracted.

Furthermore, a correction for the value of the degree on the thermometer, as was stated in the description of the thermometer, should be introduced into the calcu-

lation of molecular weight determinations by the boiling-point method, when solvents are used which boil
above 100°.

Example.— Phenanthrene in solution in benzene.[1]
$C_{14}H_{10} = 178$. In 22.95 grams benzene (22.75 used
in the calculation) gave:

Grams substance.	Rise in boiling-point.	Molecular weight.
0.1983	0.125°	182
0.6187	0.389°	182
1.0177	0.639°	183
1.6481	1.023°	185
2.2634	1.391°	187
3.0476	1.833°	191
3.9025	2.393°	187
4.6718	2.772°	194

The value 26.1, contained in the table to be given
later, is used in the calculation, as the boiling-point
constant.

The flames are extinguished after the molecular
weight determination is completed, and the apparatus
immediately taken apart. Special attention is to be
paid, in cleaning the apparatus, to the removal of
every trace of substance and solvent from the filling
material. This is collected in a funnel with narrow
neck, and washed at once with a little of the solvent.
The substance used in the investigation, can generally
be removed by evaporating the filtrate, or volatilizing
it with water vapor. The filling material is boiled
repeatedly with alcohol, or some other solvent for the
substance, and well dried before a new determination,

---

[1] From experiments by William Biltz. Next to the last experiment probably contains an experimental error.

in order that no liquid should remain unobserved in the small crevices and depressions. The cleansed and dried filling material is, after a number of determinations, warmed with concentrated sulphuric acid, then boiled repeatedly with distilled water, finally with alcohol, and then dried again.

THE BOILING-POINT APPARATUS OF JONES

A modification of the boiling-point apparatus of Beckmann has been proposed by H. C. Jones.[1] Of all the forms of boiling-point apparatus described, only those of Hite,[2] Sakurai,[3] and Landsberger[4] prevent the cold solvent from the condenser, from coming in contact with the bulb of the thermometer, before the liquid has been reheated to the boiling-point. But in no form thus far devised, has the effect of radiation outward from the hot bulb of the thermometer been properly prevented. This is accomplished by the apparatus described by Jones, which also prevents the cold, recondensed, solvent from coming in contact with the thermometer, before it is heated again to the boiling-point.

The apparatus is sketched in Fig. 35*a*. It is drawn approximately to scale. A is a glass tube 18 cm. high and 4 cm. wide. At the top it is drawn out to a diameter of about 2$\frac{3}{4}$ cm., and ground to receive a ground-glass stopper. This tube is filled to a depth of from 3 to 4 cm. with glass beads. P is a cylinder

[1] Am. Chem. J., 19, 581.

[2] *Ibid.* 17, 507.

[3] J. Chem. Soc. (London), 61, 989.

[4] Ber. d. chem. Ges., 31, 458.

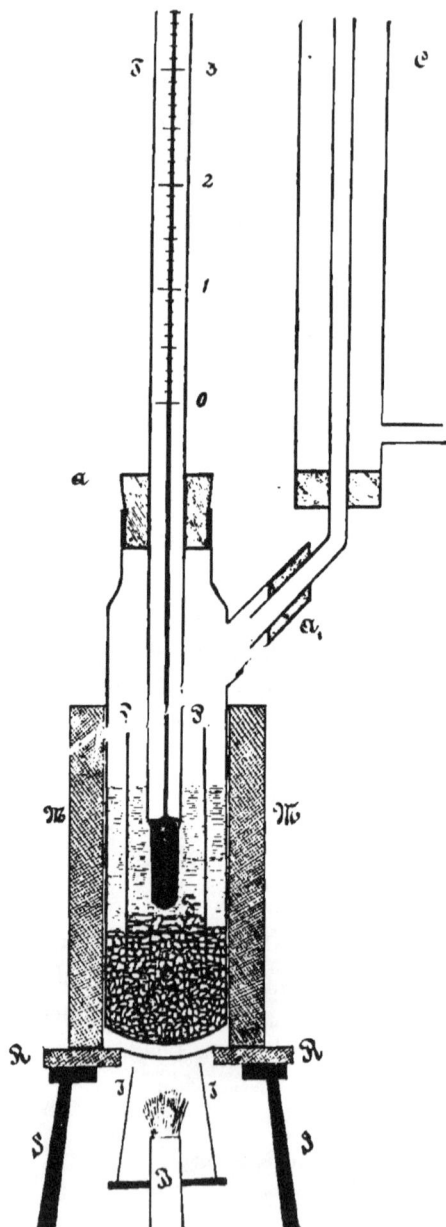

Fig. 35a.

of platinum 8 cm. high and 2½ cm. wide, made by rolling a piece of platinum foil, and fastening it in position by wrapping it near the top and bottom with platinum wire. It would be better if the edges of the cylinder were closed by welding, so that none of the liquid could pass through from one side to the other. A cylinder of some other metal, such as copper, zinc, or silver, could be employed in many cases where they would not be acted upon by the solvent or the solution, but platinum is to be preferred, because of its greater resisting power to the action of such agents. Into the cylinder P, some pieces of platinum foil are thrown. These are made by cutting foil into pieces about ¾ cm. square, bending the corners alternately up and down, to prevent them from lying too closely upon one another, and serrating the edges with scissors, to give a greater number of points from which the boiling can take place. The bulb of the thermometer is thus entirely surrounded by metal at very nearly its own temperature, except directly above. The thermometer could be forced through a hole in a sheet of platinum, which would remain suspended just over the cylinder, but in consideration of the small angle through which radiation can take place in an upward direction, this seems to be a superfluous refinement. A condenser (C), about 40 cm. long, is attached to the tube (A) which is 2 to 2½ cm. in diameter, by means of a cork. When it is desired to protect the solvent from the moisture in the air, the top of the condenser tube should be provided with a tube containing calcium chloride, or phosphorus pentoxide.

During an experiment the vessel (A), is closed by a cork, through which the Beckmann thermometer (T), passes. M is a jacket of asbestos, 12 cm. high and 1 ½ cm. thick, over the top of which the rate of boiling can be observed very satisfactorily. It is constructed by bending a thin board of asbestos tightly around the tube (A), and fixing it in place by means of copper wire. Thick asbestos paper is then wound around this, until the desired thickness is reached. The apparatus is supported on a small iron tripod (S), 8 cm. in diameter, on which rests an asbestos ring (R), about 9 cm. in external diameter. A circular hole is cut in the center of this ring, about 3 ½ cm. in diameter, and over this is placed a piece of fine copper gauze. The source of heat is a Bunsen burner (B), surrounded by an ordinary metallic cone (I), to protect the small flame from the effect of air currents. The glass vessel (A) is shoved down until it comes in contact with the wire gauze. Under these conditions, a very small flame suffices when low-boiling solvents are employed, and not a large flame is required when a solvent like aniline is used.

CARRYING OUT A DETERMINATION WITH THE JONES APPARATUS.

The glass beads are poured into the glass cylinder, the platinum cylinder inserted, and pressed down into the beads to a distance of from ½ to 1 cm. The platinum plates are then introduced into the platinum cylinder, the end of the tube (A) closed with a cork, and the ground-glass stopper inserted in A. The apparatus is then set in a small beaker glass and weighed. The solvent is then introduced, and the whole reweighed. Great care must be taken that not enough solvent is employed to boil over from one side of the platinum cylinder to the other. In case a laboratory is not provided with a balance capable of weighing accurately 200 or 300 grams, the solvent must be weighed directly, and then poured into the apparatus. This, for low-boiling solvents, is necessarily less accurate than the above-described method of procedure. After the solvent is weighed the glass stopper is removed, and the thermometer, fitted tightly into a cork, is placed in position, as shown in the drawing. The apparatus is then placed upon the stand in the mantle of asbestos, the cork removed from A, and the condenser attached. Heat is then applied and the solvent boiled. The size of the flame must be so regulated by means of a screw pinch-cock, that the boiling is quite vigorous, but not so violent as to be of an irregular or explosive character. A quiet but very active boiling is absolutely essential to the success of the experiment. The time required to establish the true temperature of equilibrium between the pure liquid solvent and its vapor, is very much greater than in the case of a solution. This is strictly analogous to what is observed with the freezing-point method. Before taking a reading on the Beckmann thermometer, it is always necessary to give it a few sharp taps with a lead pencil, and, indeed, this should be done occasionally while the mercury is rising, and especially when it is near the point of equilibrium. The use of an electric hammer to accomplish this object, is an unnecessary complication. A small hand lens, magnifying a half-dozen times, is quite sufficient to use in making the readings. It is always best to redetermine the boiling-point of the solvent. After the boiling-point of the solvent has been determined, a tube containing the substance pressed into pellets, is weighed, and a convenient number of these poured into the solvent, either through the condenser, or directly through the tube A, when the solvent is not too volatile, and has ceased to boil. The tube is then reweighed, the amount of substance introduced being thus ascertained. The boiling-point of the solution is then determined. The carrying out of a determination with a low-boiling solvent, is a much easier process than with one boiling at a considerably higher temperature.

Thus, when anisol or aniline is employed, much care and some experience are necessary to determine the rate of boiling which must be adopted. If the boiling is too slow, the thermometer will never reach the temperature of equilibrium. If so rapid that it is irregular and explosive, the thermometer may rise above the true boiling-point, and then suddenly drop below it, at the moment when a large amount of vapor is formed. In a word, for high-boiling solvents the rate of boiling must be as vigorous as possible, in order to proceed with perfect regularity.

All the forms of apparatus thus far described, are somewhat dependent upon the size of flame used. This is probably due, in part, to the corresponding change in rate at which the condensed solvent is returned to the boiling liquid. If this be true, then that form of apparatus which prevents the condensed solvent from coming in contact with the thermometer until it has been reheated, should be least affected by the size of flame used, and such is the fact.

The barometer must be very carefully noted before and after each boiling-point determination, and a correction introduced for any change in the barometric height.

This boiling-point method has been applied to the determination of molecular weights,[1] and has been found capable of yielding excellent results, with both low- and high- boiling solvents.   It has also been applied by Jones and King,[2] and subsequently, far more extensively by Jones,[3] to the measurement of electrolytic dissociation in methyl and ethyl alcohols.   This apparatus is simpler than the best forms devised by Beckmann, and eliminates experimental errors which are present in all of the modifications thus far proposed.

Modifications of the Boiling-vessel.[4] — The boiling-

[1] Am. Chem. J., **19**, 590.

[2] *Ibid.*, **19**, 753.

[3] Results will soon be published in the *American Chemical Journal*.

[4] Different modifications of the Beckmann apparatus have been described.   These appear to differ in their general applicability, and some of them must be regarded as complications.   Compare B. H. Hite : Am. Chem. J., **17**, 507 (1895); W. R. Orndorff and F. K. Cameron : Ztschr. phys. Chem., **17**, 637 (1895) ; P. Fuchs : Ztschr. phys. Chem., **22**, 72 (1897). Other modifications have been proposed by Beckmann : Ztschr. phys. Chem., **15**, 656 (1894). Another form of boiling-point apparatus, based upon the principle of the Sakurai apparatus, has recently been described by Landsberger : Ber. d. chem. Ges., **31**, 458, and very recently simplified by Walker and Lumsden, J. Chem. Soc., 502 (1898).

vessel described, nearly always suffices for the purpose
of a chemical laboratory, and is to be very highly
recommended on account of its convenient manipulation. Some changes made for definite purposes, have
been described recently by Beckmann.[1]

In case the vapor of the solvent attacks cork, the
apparatus should be closed with asbestos packing,
instead of with cork. But since this easily absorbs a
large quantity of liquid, it is better to fuse the condenser tube directly on to the boiling-vessel, thus entirely avoiding the use of the side tube. In consequence of this arrangement, no drops fall, from time to
time, from the condenser, which, as already observed,
produce small changes in the concentration. The
condensed liquid thus flows down the walls at a uniform rate, preventing the changes in concentration
and in temperature, which are produced by the introduction of a cooler drop into the boiling liquid. On
the other hand, the convenient means of determining
the amount of boiling from the rate of dropping, is no
longer available. Since the error due to the cause
first mentioned, is only slight, the tube of the condenser should be so fused into the side tube of the boiling-
vessel, that it projects somewhat into it. The drops
then fall from the protruding end, just as when the apparatus is set up simply with a cork.

It is difficult to avoid the use of a cork to close the
second opening in the boiling-vessel, where the thermometer enters. To reduce to a minimum the amount
of solvent which distils against the cork, the boiling-

[1] E. Beckmann : Ztschr. phys. Chem., 15, 666 (1894).

vessel is lengthened above the side tube, until the cork stands at about the middle of the condenser along by its side, *i. c.*, somewhat higher than is shown in drawing (Fig. 33). This lengthening is all the more possible, now, since boiling-point thermometers are recently obtainable with a very long stem between bulb and scale. The vapors in this arrangement scarcely rise up to the cork, in which the thermometer is inserted. There is another expedient which can be resorted to with substances of great activity, such as bromine. The boiling-vessel is narrowed somewhat above the attachment of the side tube, and made long enough to receive the thermometer even up to the mercury reservoir. The position of the mercury is then read through this lengthened portion of the apparatus. The apparatus is closed with a piece of rubber tubing, or with a cork in case a somewhat wider portion is attached to the narrower portion. The boiling-vessel must not be made so narrow that liquid will be drawn, by capillary attraction, well up into the space between the thermometer and the surrounding glass wall.

To prevent the boiling-vessel from cracking below, where the platinum rod is fused in, a small piece of mica or asbestos paper, with a hole cut in the center for the platinum rod, is glued around the projecting end of the rod, with a drop of water glass. The platinum rod, in this case, must protrude about 0.3 mm. from the glass. The boiling-vessel, according to my experience, does not crack at this place during use, so that any special protection appears to me to be superfluous. The vessel is much more liable to crack if, when filled with any solvent for cleansing, it is exposed to great differences in temperature. It is best to entirely avoid heating it with a free flame, and in cleansing it to warm it in a water-bath.

Some glass wool is placed on the bottom of the boiling-vessel, as a further protection, to prevent the garnets or glass beads from striking against the bottom and scratching it. This precautionary measure is not necessary when platinum is used as the filling material.

Modifications of the Boiling-jacket. — The glass boiling-jacket[1] described, can be employed under all conditions, at high temperatures as well as at low, and on account of its transparency, is to be strongly recommended for every one who wishes only to learn the method. Boiling-jackets made of porcelain are more durable. These have, in general, the form of the glass apparatus, without the bulge above and below, so that their form is exactly cylindrical, as shown in section in Fig. 36. They have a tube (a), into which the condenser is inserted, like the glass apparatus, and generally a tube is, also, placed upright above this one in the upper wall, into which a thermometer can be introduced to determine the temperature in the vapor-jacket. Two windows, through which the boiling-vessel can be observed, are cut opposite one another. (These windows are shown in the figure.)

---

[1] The glass boiling-jackets furnished by F. O. R. Götze are especially well constructed. I have had some in use for years, and have subjected them to a very wide range of temperature without injuring them.

Each of these windows is closed inside and outside, in the planes of the inner and outer walls, with a piece of mica, which is cemented on with a paste of water-glass and chalk. This can be easily accomplished, since, in the newer pieces of apparatus, a ridge projects for this purpose from the inner side, and a correspond-

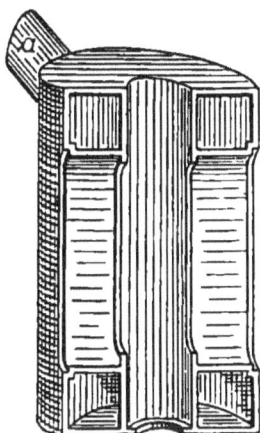

Fig. 36.
Porcelain boiling-jacket. One-third natural size.

ing groove is made in the outer side, as can be recognized on the left side of the figure. The object in closing the apparatus in this way, is to keep away currents of air which would cool the boiling-vessel. The jacket with the central opening from top to bottom, is also provided below with a narrow ledge, which can likewise be recognized in the cross section. A ring cut out of soft asbestos board, is placed upon this, and the boiling-vessel fits into the opening. The space between the boiling-vessel and boiling-jacket is thus closed below. Some fibrous asbestos, or several layers of strips of asbestos paper, are used above in the same manner, and for the same purpose. The tube of asbestos paper, which reaches from the boiling-vessel down into the ring of the heating-box, is thus dispensed with.

These porcelain boiling-jackets, like those of glass, are filled with the solvent until the level of the liquid is about half-way up the filling material in the boiling-vessel. If an upright tube is attached above to the jacket, the height of the liquid column can be easily ascertained by introducing a glass rod into it.

If the solvent attacks the cork vigorously, the condenser is attached to the boiling-jacket with some asbestos cord. When expensive solvents are employed, some other substance, having nearly the same boiling-point, is introduced into the boiling-jacket. The boiling-point of the latter can then be changed by adding another volatile substance. This is added through the condenser, until the thermometer, immersed in the vapor, registers the boiling-point desired. The boiling-points of the two substances in the vapor-jacket, should not differ more than 50°. Beckmann[1] has described other forms of boiling-jackets.

The Heating. — The heating-box is warmed by two burners placed beneath it. As already observed, these are so arranged that their flames do not directly enter the chimneys. In the case of high-boiling liquids, which require the burners to be turned on full, care must be taken that the flames do not strike under the asbestos rings of the heating-box, and fall directly

---

[1] E. Beckmann : Ztschr. phys. Chem., **15**, 666 (1894).

upon the boiling-vessel, because an irregular supply of
heat would thus result, and there would be danger of
overheating. The boiling-vessel is warmed only indirectly and, indeed, chiefly through the boiling-
jacket. The small additional amount of heat necessary to maintain the boiling, is obtained through the
tube of asbestos paper, from the closed air-chamber,
which is terminated below by the gauze of the heating-
box, and above by the boiling-jacket. Solvents having greater heat of vaporization, require more heat to
boil them. The tube of asbestos paper is then abandoned, so that the boiling-vessel projects directly into
the air-space already mentioned.

A direct heating with a flame is necessary, only
when water is used, which is distinguished by a very
high heat of vaporization, and with some very high-
boiling solvents, as nitrobenzene. A small, luminous
flame, about $\frac{1}{2}$ to $\frac{3}{4}$ cm. high, is suitable for this
purpose. It is placed far enough below the boiling-
vessel to prevent the deposit of soot upon the vessel.
A lead tube, bent in the form of an ∞, with one end
narrowed to a small opening from which the flame escapes, serves as the burner. The lead tube is supported
on a laboratory stand, placed near the boiling apparatus.
The Beckmann burner (Fig. 37) can be used instead
of this primitive device, which is, however, quite sufficient. If the burner tube is screwed off, a burner is
obtained whose use is apparent from what has been
said. The burner already set up, is used conveniently
instead of the Bunsen burner, for heating the apparatus. It is especially convenient with very high-

boiling solvents, since three or four such burners can be placed under the heating-box, while the large bases of Bunsen burners are in the way.   A circular burner, which gives about 35 small flames, can be attached to these burners furnished by the mechanic, J. G. Böhner of the Physical Institute in Erlangen, by means of a small porcelain tube and asbestos packing.   When the

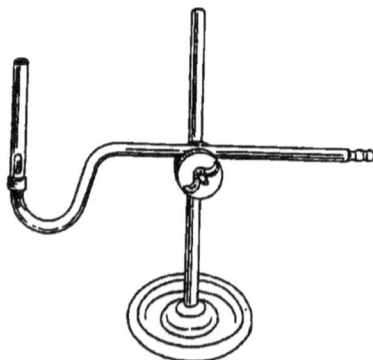

Fig. 37.
Adjustable burner. One-sixth natural size.

burner is used, the circle of burners is placed some-what below the outer asbestos ring of the heating-box. Higher iron stands are necessary for the heating-boxes, that these burners may be used conveniently.   They can be obtained 25 cm. high.

Marked changes in the gas-pressure in the city mains, which usually take place toward evening, can sometimes do harm.   The introduction of a gas-regu-lator makes us independent of the pressure of the gas in the conduit tube.   Beckmann recommends the use of the one attached to the Gulcher thermopile made by the firm of Julius Pintsch in Berlin.   Smaller variations, which sometime occur during the day, have no appreciable effect.

The Introduction of the Substance. — Solid substances are compressed into tablets, exactly as described under the freezing-point method. Generally, the tablets can be fairly tightly compressed, since they are usually dissolved easily and quickly in the boiling solvent. When the substances are more difficultly soluble, the tablets are, of course, not compressed so tightly, so that they break up in dissolving ; or small vessels of fine platinum gauze are prepared, filled with the substance in the form of powder, and closed above by bending the walls together. Glass, or metallic vessels cannot be used, because the substance is dissolved out of them only very slowly. A layer of concentrated solution forms over the substance lying on the bottom of the vessel, which prevents or delays further solution. Small flat boxes, or boats of thin platinum foil, have occasionally proved to be satisfactory. Small conical glass tubes, open below, can also be employed. These remain suspended between the thermometer and the wall of the boiling-vessel, their contents being washed out by the condensed liquid as it flows down.[1] But tablets are to be used whenever it is possible, that the height of the column of liquid should not be unnecessarily increased through the introduction of foreign substances. A rise in boiling-point accompanies a rise in the liquid column, which, although it is not great, yet, whenever possible, should be avoided.

---

[1] C. Schall : Ztschr. phys. Chem., 12, 147 (1893).

According to the statements of Beckmann[1], the rise in the boiling-point of ether, under 760 mm. pressure, is 0.002° for each millimeter increase in the height of the layer of ether.

Liquids are introduced with a weighing pipette, which is quite similar to that used in cryoscopic determinations, but is to be distinguished from it in that the capillary is much longer, indeed, somewhat longer than the condenser. In introducing the liquid, the pipette is shoved into the condenser until the end of the exit tube is exactly at the place where the vapor of the solvent condenses. Only such a small trace of the solvent condenses, then, on the point of the pipette, that it can be neglected. But it is always desirable not to allow the liquid contained in the capillary of the pipette to flow back, until the pipette has been raised into the upper portion of the condenser, so that none of the solvent will be drawn into the pipette. The following procedure can also be adopted. The desired amount of liquid can be dropped into the condenser, just above the point of condensation of the vapors, when no trace of the solvent can condense upon the pipette, and then about ½ cc. of the solvent can be dropped in from a second pipette, to rinse down the substance. The amount of solvent introduced, can be determined by weighing the second pipette, and this is added to the amount already in the apparatus.

The boiling-point method is, in general, rarely employed for liquids, since the freezing-point method is better adapted to them, and, indeed, a suitable solvent

---

[1] E. Beckmann : Ztschr. phys. Chem., **4,** 549 (1889).

can always be found when it is to be used. Very vis-
cous liquids are most simply introduced into the boiling apparatus by means of a small box made of platinum foil.

The Use of the Different Solvents. — The boiling-
point method has the advantage over the freezing-point,
that the different solvents can be employed with it
without any essential change in the manner of using
the apparatus; while with the freezing-point method,
different devices are necessary for keeping the temperature of the thermostat constant. In addition to the
high-boiling solvents, water, as already observed, is
the only exception. It requires a direct heating of
the boiling-vessel with a flame, since its heat of vaporization is enormously high.

A thick layer of fine-grained filling material, say garnets, is introduced into the boiling-vessel, when
solvents are used with large heats of vaporizations, in
order that the superheated bubbles of vapor, rising
from below, should be very often checked in their
path, slowly turned aside, and given time and abun-
dant opportunity to part with the excess of heat. They
would then not reach the layer of liquid surrounding
the thermometer, until all superheating had been done
away with. A layer of garnets from 4 to 5 cm. in
thickness, should be employed with water as a solvent.
On the other hand, a layer of glass beads, from 3 to
3½ cm. in thickness, suffices for solvents whose heat of
vaporization is small.

Hygroscopic solvents, *c. g.*, ethyl acetate and acetic
acid, are protected from the outer air by a small tube

about 5 cm. long, filled with calcium chloride. This is attached to the condenser, with a cork, and is removed for a moment while the substance is being introduced.

Ethyl ether is an especially convenient solvent. It can be easily obtained of sufficient purity, readily dissolves many substances, boils at a convenient temperature, so that generally a constant temperature is quickly reached, gives results which are simple to interpret, and permits the substance used to be easily recovered. .

It is more difficult to carry out molecular weight determinations correctly with water, since the elevations obtained with dilute solution are very small, because of the small value of its boiling-point constant. Satisfactory values can, however, be obtained with it, especially at medium concentrations, by working carefully and slowly.

A *careful purification* of the solvent to be used, is of the greatest importance for the boiling-point method. Special directions for the purification of the several solvents have been given by Beckmann, Fuchs, and Gerhardt.[1] It generally suffices to purify the liquids by the usual chemical methods, to carefully dry them, and to distil them, using a fractionating apparatus. The substance to be employed, must pass over within a few tenths of a degree. The pure preparation should be kept in glass flasks with tightly fitting glass stoppers, or still better, in small pipettes whose ends

---

[1] E. Beckmann, G. Fuchs, V. Gerhardt: Ztschr. phys. Chem., **18,** 496 and following (1895).

are fused together, of the form of the Ostwald pycnometer.[1] Many solvents, such as aniline, ethylene bromide, the ethyl compounds of the halogens, chloroform, some ethereal salts, decompose easily in the light, and are therefore kept in the dark.

	Boiling-point.	Molecular heat of vaporization [2]	Boiling-point constant.
Acetone	56°	125	17.1
Acetonitrile[3]	81°	139	17.9
Ethyl acetate[4]	77°	90	26.8
Ethyl ether[5]	35°	87	21.6
Ethyl alcohol	78°	208	11.7
Ethyl bromide	38°	69	27.9
Ethylene bromide	130°	50	64.5
Ethylene chloride	83°	81	30.9
Ethyl formate	54°	100	21.2
Ethylidene chloride	57°	69	31.3
Ethyl iodide	72°	46	51.6
Ethyl mercaptan	37°	100	19.0
Ethyl sulphide[6]	90°	80	32.6
Amyl alcohol (Iso)	131°	125	25.8

[1] E. Beckmann : *Ibid.*, **21**, 251 (1896).

[2] The heats of vaporization have been calculated by means of the formula already given, from the boiling-point constant and the absolute boiling temperature.

[3] Acetonitrile undergoes slow change when boiled, and therefore does not give a sharp boiling-point.

[4] It is best to purify ethyl acetate just before the experiment, by shaking it repeatedly with water, carefully drying it, and then distilling it, since older preparations usually contain some alcohol and acetic acid. The same holds for methyl acetate. Since these ethereal salts are very hygroscopic, the condenser must be closed with a calcium chloride tube.

[5] Ethyl ether is to be shaken with mercury, after it is cleaned and dried, thus removing a product which raises the boiling-point. W. Ramsay and J. Shields : Ztschr. phys Chem., **12**, 448 (1893).

[6] Ethyl sulphide is adapted as a solvent for many inorganic salts. Compare M. Stephani : Dissociation Zürich., p. 29 (1896).

	Boiling-point.	Molecular heat of vaporization.	Boiling-point constant.
Amylene hydrate (tertiary amyl alcohol)	102°	113	24.6
Aniline	184°	129	32.0
Anisol	155°	92	44.3
Benzene	79°	94	26.1
Benzonitrile	191°	117	36.5
Bromine	58°	46	47
Chloroform	61°	61	35.9
Cymol	173°	71	55.2
Diethyl sulphide	92°	66	40
Dipropylamine	106°	62	46.0
Acetic acid[1]	118°	120	25.3
Isoamyl acetate	142°	71	48.3
Isobutyl alcohol	108°	143	20.1
Isopropyl alcohol	83°	194	12.9
Camphor	204°	77	58.5
Menthol	212°	71	65.2
Menthone	206°	73	62.5
Methyl acetate	56°	104	20.6
Methylal	42°	93	21.1
Methyl alcohol	66°	259	8.8
Methyl formate	32°	116	15.8
Methyl iodide	42°	46	42.3
Methyl-propyl ketone	102°	92	30.3
Nitroethane	114°	116	25.5
Nitrobenzene[2]	209°	92	50.0
Paraldehyde[3]	123°	74	41.8
Phenol	183°	135	30.4
Propionitrile[4]	97°	120	22.6

[1] Acetic acid, when boiled for a long time, strongly attacks the corks.

[2] The constant for nitrobenzene is calculated from the best determinations available. Ztschr. phys. Chem., **19**, 424 (1895).

[3] Paraldehyde, on boiling, partly passes over into acetaldehyde, so that the boiling-point is difficult to determine: Ztschr. phys. Chem., **18**, 507 (1895).

[4] Nitriles of the fatty series undergo change on boiling, and therefore do not give a boiling-point which remains constant for any length of time. Compare Werner: Ztschr. anorg. Chem., **15**, 33 (1897).

	Boiling-point.	Molecular heat of vaporization.	Boiling-point constant
Propyl alcohol (normal)	97	170	15.9
Mercury	357	60	130.0
Carbon bisulphide	46°	86	23.5
Carbon tetrachloride	76°	50	48.0
Water	100	540	5.1

Some Solvents Which Cannot be Used in Certain Cases.

— All solvents cannot be indiscriminately used with the boiling-point method, just as with the freezing-point method, and indeed for a very similar reason. As in the latter case, a mixture of solvent and the dissolved substance sometimes separates on solidification, so it can happen here, that the dissolved substance also volatilizes with the vapor of the solvent. The concentration of the solution remaining behind, is therefore smaller, and the molecular weights found are too large.

A correction can be made, as Beckmann and Stock[1] have shown, on the assumption that the dissolved substance has the same molecular weight in the gaseous and liquid states of aggregation, by multiplying the result found for the molecular weight M', leaving out a correction, with the expression $(1 - a)$. a is the relation of the concentration (*i. e.*, the number of grams of dissolved substance to 100 grams solvent) in the space occupied by the vapor, to the concentration in the liquid solution. This relation, from Henry's law, is constant for solutions of any concentration of the same substance in the same solvent. It is ascertained as follows : A solution of known concentration of the

---

[1] E. Beckmann and A. Stock : Ztschr. phys. Chem., **17,** 110 (1895).

substance to be investigated, in the solvent chosen, is distilled, taking the precaution that the vapors are condensed only in the condenser. The distillation is interrupted after a time, and the amount of the distillate and of the dissolved substance contained in it, are determined analytically. The concentration of the distillate, which is equal to that in the space occupied by the vapor g_1, is calculated from this. The original concentration of the boiling solution is known. The final concentration, which obtains after the first portion is distilled off, can be calculated from the amounts of solvent and dissolved substance which have passed over. Let g_2 be the mean of the concentrations at the beginning and end; i. e., the mean concentration of the boiling solution. The relation between the concentration of the portion which has distilled over, and that of the distilling solution, is then: $\frac{g_1}{g_2} = a$. Several values for a, the mean of which can be used for correcting the molecular weight, can be obtained, by distilling over after the first portion, a second portion, etc., and investigating it.

This method of introducing correction is simple, only when the fractions can be easily analyzed, as by titration; or in case only one of the two substances to be taken into account contains nitrogen or a halogen, by an analytical determination of these elements. It is not practicable when analytical determinations present difficulties, e. g., when both substances are hydrocarbons.

The investigation of the molecular weight of iodine,

by Beckmann and Stock, serves as an example. Iodine sublimes at $110°$ to $120°$; carbon tetrachloride boils at $76.5°$.

DETERMINATION OF a FOR IODINE IN CARBON TETRACHLORIDE.

	1	2	3
Number of the fraction			
Concentration of the distillate g_1	0.579	0.732	0.899
Mean concentrations of the boiling solutions g_2	1.656	1.947	2.348
a	0.35	0.38	0.38

Mean value of a from all the experiments is 0.37.

The corrected molecular weight M, is obtained from the uncorrected molecular weight M_1, by introducing this value into the above formula.

UNCORRECTED AND CORRECTED MOLECULAR WEIGHTS OF IODINE IN CARBON TETRACHLORIDE.

M_1	370	365	374	382
M	233	230	236	241

Calculated I_2 254.

The value of a is smaller the greater the difference between the boiling-points of solvent and of dissolved substance. Beckmann and Stock found the value of a to be 0.1, when methylal, whose boiling-point is about $42°$, was used as solvent for iodine. The introduction of this value diminished only slightly the value of M_1. If the dissolved substance boils more than $130°$ higher than the solvent, the effect of the part carried along with the vapor, is so small, that it no longer needs to be taken into account. Only in this case can the molecular weight be derived from the results of experiment, without further correction. *To avoid the troublesome correction, always choose a solvent whose boiling-point lies at least 130° below that of the dissolved substance.*

A corresponding correction can be introduced, also, for substances which have a different molecular weight in the gaseous form than when dissolved in a liquid. The theoretical foundations for this have been furnished by Nernst.[1] That the molecular weight of a substance in the form of gas is the same as in solution in a liquid, can be shown, on the one hand, by the fact that the uncorrected molecular weights obtained by the boiling-point method, are found to be constant within certain narrow limits, independent of the concentration, which would not be the case if the molecular weights were different in the two states of aggregation. It is further shown, in that the value of a, determined experimentally, is always found to be the same, independent of the concentration of the solution.

Effect of Atmospheric Pressure. — The boiling-point of a liquid is dependent upon the pressure of the atmosphere resting upon it. Annoying disturbances can be produced in the exact determinations of the boiling-point, by the Beckmann method, by changes in the height of the barometer. An idea of the magnitude of these can be obtained, by noting that a change in pressure of 1 mm. corresponds to a change in the boiling-point of 0.03° to 0.04°.[2] A change in pressure of several millimeters can occur within a few hours, on stormy days.

---

[1] W. Nernst : Ztschr. phys. Chem., **8**, 128 (1891).

[2] H. Landolt and R. Börnstein : Physikalisch-chemische Tabellen, 2nd Edition, Tables 25 to 37.

It is difficult to remedy this. One way out of the difficulty is to measure the pressure on an exact barometer from time to time, during the molecular weight determination, and to take the corresponding changes in boiling-point from the tables at hand.[1] The boiling-points, read on the thermometer, are then corrected by these amounts. This correction can be applied only with the more common solvents, for which the corresponding tables are available.

The exact reading of the barometer, which generally is not practicable in the chemical laboratory, can be avoided by observing the changes in the boiling-point of water with an accurate thermometer, and taking from a table the changes in pressure corresponding to these.

An attempt has also been made to introduce a correction by allowing the pure solvent to boil in a second apparatus, and to use the changes in the boiling-point as correction values. But this method does not appear to be very reliable.

Although the effect of changes in the barometer can be so considerable, yet no serious disturbances are thus produced in simple molecular weight determinations, as carried out in the laboratory. Since, in the short time required to make some molecular weight determinations, after the boiling is constant (I allowed ten minutes for each experiment, Beckmann a still shorter time[2]), the pressure changes so little that such changes

---

[1] H. Landolt and R. Börnstein : Physikalisch-chemische Tabellen, 2nd Edition, Tables 25 to 37.

[2] E. Beckmann : Ztschr. phys. Chem., **15,** 676 (1894).

do not need to be considered.[1] When the concentration of the solution has become greater after the first experiment, small changes are no longer of much consequence. It is, however, always desirable to read the barometer before and after the determinations, and to repeat the experiment, if the determinations show a difference sufficient to produce an error of more than five per cent. in the result. Resort must be had to the methods of introducing correction, just mentioned, only when an accurate study of dissociation curves is being made, and when it is therefore necessary to carry out, in succession, a longer series of molecular weight determinations.

THE RESULTS

1. **Smaller Deviations Inherent in the Method.**— Influences on the results manifest themselves in the boiling-point method, quite similar to those in the freezing-point, only their action is different. The influences investigated by Noyes,[2] which depend upon the space occupied by the molecules of the solvent and of the dissolved substance, have the same effect in both methods, because through these the osmotic pressure, which is determined indirectly by both methods, is changed. Their action is to give higher

[1] Translator's note: Our experience in determining molecular weights and measuring dissociation by the boiling-point method, in this country, has shown the necessity of the closest attention to changes in the barometer, correction being necessary in many experiments which lasted only a few minutes. It seems, from observations, that the daily barometric change is, on the average, much greater here than in Germany.

[2] A. A. Noyes : Ztschr. phys. Chem., **5**, 53 (1890).

values for the molecular weight with increasing concentration.

An increase in the value of the constant K′, which accompanies an elevation of the boiling-point, has the opposite effect. This is explained through an elevation of the absolute boiling temperature T, and a diminution in the heat of vaporization w_1, and is expressed in the formula already mentioned.

$$K_1 = \frac{0.0198 T_1^2}{w_1}.$$

The more concentrated the solution the higher the boiling-point, and therefore the higher the true value of K_1. If only the one value of K_1, determined for the boiling temperature of the pure solvent itself, is employed instead of this, in the calculation, the values of the molecular weight at greater concentrations will be too small.

These two influences have the opposite effect. And it often happens in boiling-point determinations, that

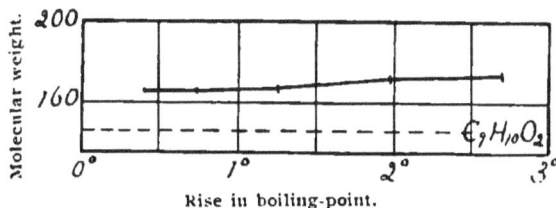

Rise in boiling-point.
Fig. 38.
Ethyl benzoate in benzene.[1]

the same, or very nearly the same, values are found for the molecular weights, within a wide range of concentrations. The curve drawn from the results of

_____
[1] E. Beckmann : Ztschr. phys. Chem., **6**, 439 (1890).

the boiling-point method, in the same manner as the curve from the results of the freezing-point method, is, therefore, often a straight line nearly parallel to the axis of the abscissae.

In many other cases, however, rising and sometimes

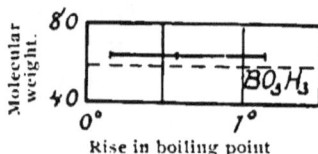

Rise in boiling point

Fig. 39.

Boric acid in water.[1]

falling curves are found, their direction being affected by the above-named and other influences. Such a deviation is shown slightly by the curve for ethyl benzoate, more strongly by the curve for phenanthrene.

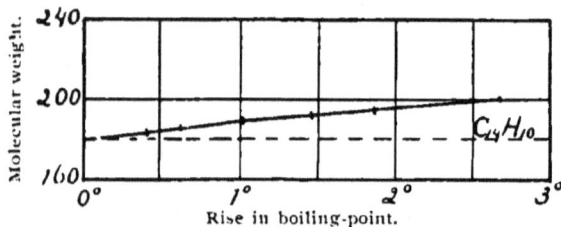

Rise in boiling-point.

Fig. 40

Phenanthrene in solution in benzene.[2]

The molecular weight at infinite dilution, can be derived from the boiling-point curve, just as from the freezing-point curve, and, indeed, generally with success. Many examples of this are given in the papers of Beckmann.[3] (Compare also curves Fig. 42.)

---

[1] E. Beckmann : Ztschr. phys. Chem., **8**, 227 (1891).

[2] From the results cited on page 160.

[3] E. Beckmann : Ztschr. phys. Chem., **8**, 227 (1891).

2. **Electrolytic Dissociation.**— The boiling-point method, like the freezing-point, shows electrolytic dissociation of electrolytes in aqueous solution, or in solution in formic acid, or methyl alcohol. The amount of this dissociation varies with the nature of the dissolved substance, and of the solvent. An investigation of sodium chloride in water showed complete dissociation. NaCl = 58.4.

In 27.69 grams water (27.39 in the calculation) gave:

SODIUM CHLORIDE IN WATER. BOILING-POINT METHOD.

Grams substance.	Rise.	Molecular weight.
0.1452	0.098^C	27.6
0.4460	0.288^C	28.8
0.8171	0.528^C	28.8
1.4347	0.920^C	29.0
2.2406	1.481^C	28.2

The value, calculated on the assumption of complete dissociation, is 29.2.

A decreasing dissociation with increasing concentration, is shown from the following determinations with potassium nitrate. KNO_3 = 101.

In 25.90 grams of water (25.55 in the calculation) gave:

POTASSIUM NITRATE IN WATER. BOILING-POINT METHOD.

Grams substance.	Rise.	Molecular weight.
0.2062	0.085^C	48.4
0.4910	0.183^C	53.6
0.9175	0.334^C	54.6
1.7774	0.615°	57.7
2.2430	0.755°	58.0
3.2870	1.055°	62.2
4.3091	1.365^C	63.0

Normal molecular weights are obtained with electrolytes in other solvents, e. g., ethyl alcohol, as the investigations of Raoult[1] have shown for a number of salts, such as potassium acetate, lithium chloride, calcium chloride, calcium nitrate. Normal values for the molecular weight also of zinc chloride,[2] are obtained in solution in ethyl alcohol.

In 18.56 grams of ethyl alcohol (18.37 in the calculation) gave :

ZINC CALORIDE IN ETHYL ALCOHOL. BOILING-POINT METHOD.

Grams substance.	Rise.	Molecular weight.
0.1959	0.087°	143
0.4096	0.194°	134
0.6825	0.336°	129

Other inorganic salts, the determination of whose molecular weight in solution would be of considerable theoretical interest, e. g., ferric chloride and aluminum chloride,[3] have, on the contrary, not been successfully investigated thus far by the boiling-point method, in organic solvents, because they act chemically upon the solvent.   This is shown by the quick rise in temperature, which immediately follows the introduction of the salt into the boiling solvent.   If there were no reaction, the thermometer would have first fallen ; therefore, this is an indication that reaction has taken place.   It is not permissible to draw a conclusion in reference to the molecular weight in question from

---

[1] F. M. Raoult : Compt. rend., 107, 442 (1888).

[2] From experiments by O. Klosmann and E. König in Greifswald.

[3] P. Th. Muller : Compt. rend., 118, 644 (1894).

the values found,[1] which, as a matter of fact, agree
with the formula FeCl$_3$. I state these details to show
how careful we must be in the critical examination of
molecular weight determinations of inorganic substances, which are investigated in organic solvents.

Recently, a comprehensive investigation has been
published by Werner[2], and his co-workers, on the
molecular weight of inorganic salts in organic solvents. In the majority of cases, the simple molecular
weights were found. The same was true for stannous
chloride, aluminum chloride, and ferric chloride. This
result for these three salts was, indeed, surprising,
since careful vapor-density determinations had shown
the existence of more complex molecules, and the highest molecular weight given by a vapor-density determination is always found in solution. Since, however, both
of the solvents used by Werner always appear to form
addition-products of the solvent, with metallic salts (and
Werner himself has furnished many instances of this),
I believe that more complicated processes take place
in the solutions, and that from his results alone, con-
clusions must not be drawn as to the size of the
molecules of the pure substance. Just as when ferric
chloride is dissolved in alcohol and acetic ether, no
reaction manifests itself by the evolution of heat, so,
according to Werner's statement, is no heat evolved
when aluminum chloride is dissolved in pyridine.

3. **Complex Molecules.**—The intimate relationship

---

[1] From experiments by O. Klosmann and E. König in Greifswald.

[2] Alf. Werner : Ztschr. anorg. Chem., **15,** 1 (1897).

between the freezing-point and boiling-point methods, which has been repeatedly emphasized, holds, finally, for those substances which tend to form complex molecules, and for those solvents which favor such an association. Both methods lead, in such cases, to the same results. The difference in temperature makes a slight difference in the results. The curve obtained by the boiling-point method,— other conditions being the same,— is lower than the curve obtained by the freezing-point method, for dissociating substances, or such which contain more complex molecules in concentrated than in dilute solutions, since the dissociation has proceeded further, or the association not as far, at the higher temperature.   (Compare Fig. 42.)

These relations have, up to the present, been worked out thoroughly only by means of the freezing-point method. The analogous investigations for the boiling-point method, which would undoubtedly lead to the same results, have not yet been carried out, so that the statements to be made here must be very incomplete.

Of the solvents named, the following favor the formation of complex molecules:

ASSOCIATING SOLVENTS.

Ethyl bromide,	Methyl iodide,
Ethylene bromide,	Methyl-propyl ketone,
Ethylene chloride,	Nitroethane,
Ethylidene chloride,	Nitrobenzene,
Ethyl iodide,	Propionitrile,
Benzene,	Carbon bisulphide,
Chloroform,	Carbon tetrachloride.
Cymol,	

On the contrary, simple molecules are found in the following solvents:

Acetone,	Isobutyl alcohol,
Ethyl acetate,	Isopropyl alcohol,
Ethyl ether,	Methyl alcohol,
Ethyl alcohol,	Methyl acetate,
Ethyl formate,	Methylal,
Amyl alcohol,	Methyl formate,
Acetic acid,	Phenol,
Isoamyl acetate,	Propyl alcohol,
Isoamyl alcohol,	Water.

In general, all substances corresponding in their constitution to the water type, and also acetone, exert a dissociating influence.

The *chemical nature* of the dissolved substances, and the *concentration*, so far as investigations up to the present have shown, produce exactly the same effect on the boiling-point curve, as on the freezing-point. A few curves will serve to make this clear.

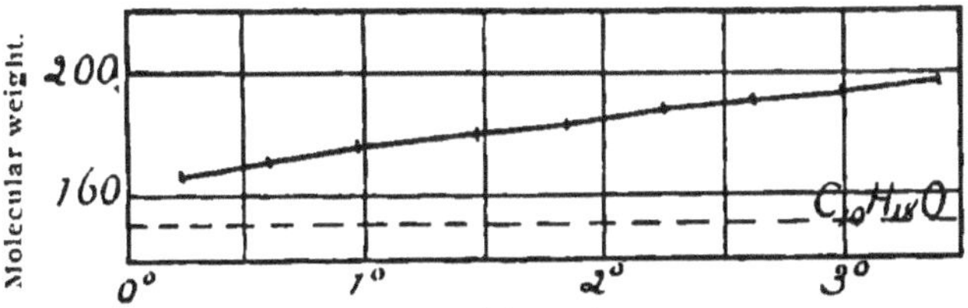

Rise in boiling-point.

Fig. 41.
Borneol in solution in benzene.[1]

The boiling-point curve of benzoic acid in benzene, is not drawn from the elevations of the boiling-point,

[1] From experiments by William Biltz.

but from the concentrations, since it should be compared with the freezing-point curve, which is introduced
above it in the drawing. It is seen that the dissocia-

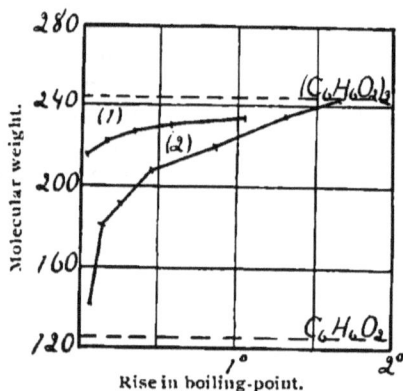

Rise in boiling-point.

Fig. 42.

Benzoic acid in solution in benzene : (1) According to the freezing-point.
(2) According to the boiling-point method.[1]

tion for equal concentrations has proceeded further
at 80° than at 5°, as would be expected.

This boiling-point curve is, moreover, especially
adapted to show how, in many cases, the value for infinite dilution can be derived very successfully by extrapolation.

The curves for borneol and *o*-formtoluide, are typical of alcohol and anilide curves, respectively.

Benzophenone in solution in glacial acetic acid, investigated by the boiling-point method, behaves
irregularly, just as an investigation by the freezing-
point method had shown, the curve rising rapidly
with increasing concentration. Benzoic acid anhy-

---

[1] Boiling-point curve from experiments by William Biltz.

dride cannot be investigated in glacial acetic acid by the boiling-point method, since it is unstable in boiling glacial acetic acid.

Benzil, phenylbenzoate, and ethyl benzoate[1], in concentrated solutions, show an increase in the molecular weight, which, though not large, is unexpected.

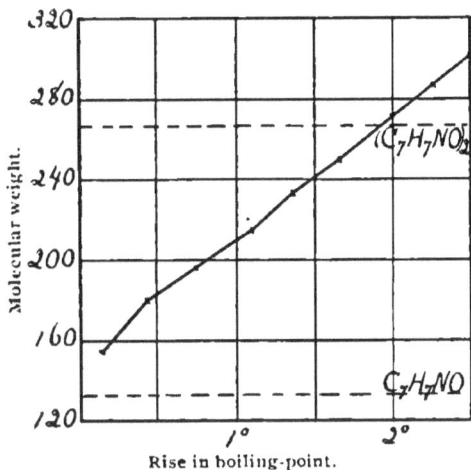

Rise in boiling-point.

Fig. 43.

o-Formtoluide in solution in benzene.[2]

Some inorganic salts also show a tendency to form double molecules, according to the investigations of Werner; e. g., cuprous cyanide, and to a less degree, cuprous bromide in solution in pyridine. Further, silver iodide in pyridine. While values were found for silver chloride and silver bromide, which were even higher than would correspond to the double molecules.

---

[1] E. Beckmann : Ztschr. phys. Chem., 6, 451 (1890).

[2] From experiments by G. Preuner.

Choice of Method. — In deciding whether to use the freezing-point, or the boiling-point method, in a given case, the solubility of the substance to be investigated must first of all be taken into account. If it dissolves easily in one of the solvents used with the freezing-point method, this method is preferable, because it gives more accurate results than the boiling-point method with smaller amounts of substance. Since the freezing-point constants are larger than the boiling-point constants, the difference in temperature, determined for equal amounts of substance, is greater when the freezing-point method is used, and errors in the measurement of temperature have less effect.

But in very many cases the solubility of the substance to be investigated, is too small to permit the use of the freezing-point method. The boiling-point method can be used in such cases, and is of incalculable value.

The carrying out of a molecular weight determination by the boiling-point method is, without doubt, more convenient than a freezing-point investigation, in case the apparatus is ready. The apparatus, when heated, can be left to itself, and then it is only necessary to observe the temperature from time to time. The determination proper, consists in introducing the substance and observing the rise in temperature, and is over in a short time. The substance dissolves rapidly and spontaneously, while this often presents difficulties in the freezing-point method. The danger also of a substance crystallizing out, which sometimes cannot be foreseen, is avoided.

The freezing-point method, as already observed, is to be unconditionally preferred in the investigation of liquids.

DETERMINATION OF MOLECULAR WEIGHTS FROM THE PRINCIPLE OF LOWERING OF SOLUBILITY

It has already been mentioned in the introduction to the chapter on osmotic methods, that a solution can dissolve less of a second liquid than the pure solvent; e. g., an ethereal solution of anthracene dissolves less water than pure ether. The law underlying this, which led to a method for determining the molecular weight of the dissolved substance, was discovered by Nernst.[1] The relative lowering of the capacity to dissolve a second liquid, which a solvent experiences on adding a foreign substance, is equal to the relation between the number of molecules of the foreign substance dissolved, and the number of molecules of the solvent.

Let L_0 be the solubility of the liquid in the pure solvent; L the solubility in solution; the relative lowering of the solubility, therefore, $\dfrac{L_0 - L}{L}$. Further, let n be the number of molecules of the substance dissolved, and N the number of molecules of the solvent; the above law can then be expressed in the following formula:

$$\frac{L_0 - L}{L} = \frac{n}{N}.$$

Further, let $n = \dfrac{g}{m}$, where g represents the amount

---

[1] W. Nernst: Ztschr. phys. Chem., **6**, 16 (1890).

of substance dissolved, and m its molecular weight. By introducing this expression for n, and proper transformation,[1] we finally obtain the equation:

$$\frac{L_0 - L}{g} m = \frac{L_0}{N + n}.$$

Since n, with respect to N, is very small, it can be neglected. It follows, then, that the right side of the equation is a constant, when an equal quantity of the same solvent (N therefore constant) is taken in every experiment. If this constant is represented by C, we have

$$\frac{L_0 - L}{g} m = C.$$

The constant C, from this equation, can be determined experimentally, by determining the lowering of solubility $L_0 - L$, which a constant quantity of a solvent undergoes, with respect to a second liquid, when g grams of a substance of known molecular weight m, are dissolved in it. It should be observed that the value of C, thus determined, is not a general constant for the solvent in question, but refers to the amount of solvent employed in the experiment, and to the temperature which obtains.

This procedure has been worked out for the practice of the laboratory, by Tollaczko,[1] a pupil of Nernst. Ether is used as the solvent, and water as the second liquid. The method can be used for all substances which are easily soluble in ether, but not soluble in water. The solubility capacity of the ether is ascer-

[1] St. Tollaczko : Ber. d. chem. Ges., 28, 804 (1895).

tained, simply by volume measurement. The following simple apparatus is employed for this purpose. A
flask of about 100 cc. capacity, is provided with a neck
about 15 cm. long, and 0.7 to 0.8 cm. wide, which is
graduated to half-millimeters. The neck can be closed
ether-tight, by means of an accurately ground-glass
stopper.

To determine C, the flask is filled up to where the
neck begins, with water saturated with ether. Then
a quantity of ether determined once for all, say a layer
of 6 cm. thick, which in turn is saturated with water,
is poured in, and finally some drops of mercury added,
so that the boundary of water and ether comes on the
lower portion of the scale. It is desirable to employ
in every case, not only the same quantity of ether, but
also the same quantity of water. After the flask has
stood for some time at a constant temperature, being
meanwhile repeatedly shaken, the mercury effecting
better mixing, the position of the bounding layer between ether and water is read exactly. A check
reading is made after a time.

A weighed amount of substance is now introduced
into the ether; 0.05 to 0.15 gram, when the substance
has a small molecular weight, and 0.1 to 0.3 gram
when the molecular weight is large. The substance
is dissolved by shaking, and the new position of the
ether accurately determined by a number of readings.
The bounding layer has now risen higher into the
neck. The difference between the two readings is the
diminution in solubility, expressed in divisions of the
scale = $L_0 - L$. C is obtained by introducing the
value into the formula.

Example: 0.0655 gram benzene produce a displacement of 0.45 cm.; the molecular weight of benzene is 78.

$$C \quad 78 \frac{0.45}{0.0655} \quad 536.$$

After the value of C has been accurately established, by repeated experiments, it can be used for determining the molecular weight of substances whose molecular weight is unknown. For the calculation, the equation used above is solved for m :

$$m \quad \frac{Cg}{L_0 - L}.$$

The molecular weight determination proper, is carried out in exactly the same manner as the determination of C.

An example may serve to make the process clear :

Naphthalene in ether. 0.1266 gram naphthalene depresses the solubility of water in ether about 0.55.

$$m \quad \frac{536 \times 0.1266}{0.55} - 123.$$

M, calculated from the formula $C_{10}H_8$, $= 128$.

This simple method, which can be easily carried out, is very useful for substances which are easily soluble in ether, but which are insoluble in water. It is to be especially recommended for the investigation of liquids. Its accuracy is certainly not very great, since the displacement of the bounding layer between the water and ether is only slight.

The fundamental condition for obtaining useful values, is that the temperature must be maintained exactly constant in all of the readings, as well in determining the constants as in the molecular weight determinations proper. The same temperature must obtain in both cases, to a tenth of a degree. A large waterbath, whose temperature is read with a thermometer graduated to tenths of a degree, is employed as a thermostat.

DETERMINATION OF THE MOLECULAR WEIGHT OF HOMOGENEOUS SOLIDS OR LIQUIDS

In the methods thus far described, the molecules were separated for the purpose of determining molecular weights, either by vaporization, or solution in a solvent. The determination of the molecular weight of undiluted solids or liquids, has been recently accomplished, a problem, whose solution a short time ago would have appeared scarcely possible.

These methods, up to the present, have been chiefly worked out with homogeneous liquids, and solids since they are less simply dealt with, have been investigated in only a few cases. It has been shown that the majority of organic substances, in the solid and liquid conditions, are made up of the corresponding simple molecules. Double, and still more complex, molecules exist in those substances, in which, the investigation of their solutions had pointed to a tendency towards the formation of such molecular complexes, as with the acids and alcohols. More complex molecules exist, also, in some substances in the homogeneous solid, or liquid condition, whose presence was not indicated from the investigation of their vapor or solutions ; thus, solid naphthalene consists of double molecules.

The existence of molecular complexes is shown, indeed, by some substances in the form of vapor ; c. g., by the simpler monobasic fatty acids. The same peculiarity is shown by a larger number of classes of substances in the state of liquid solution. And the same is shown by still other substances, also in the homogeneous condition.

The assumption formerly made, that solid substances consist of larger aggregations of the simplest molecules, and therefore always have a much greater molecular weight than the substance in the liquid or gaseous condition, is thus shown not to be well founded.

The problem of determining the molecular weight of a homogeneous substance, is solved by different methods. The majority of them give the " association factor ;" *i. e.*, the number of simple molecules which are united to form a complex molecule. The association factor is often a fraction; *viz.*, when the substances are in a state of dissociation of the more complex molecules into simpler molecules, simple molecules also being present, together with more com. plex molecules. The association factor for acetic acid[1] at 200°, is about 1.5, while at 50° it is 2. Some of the double molecules, which are stable at 50°, are decomposed, therefore, at the higher temperature. More detailed information in reference to the individual methods for determining the association factor, can be found in the paper by Ramsay, already referred to, and in a summary of the different methods by Traube.[2]

The simplest method for determining the associa-

---

[1] W. Ramsay : Ztschr. phys. Chem., **15,** 111 (1894).

[2] J. Traube : Ber. d. chem. Ges., **30,** 265 (1897).

tion factor of homogeneous solids or liquids, has been discovered by J. Traube, a method which shows at once the formula of the molecule. This method differs from those thus far described for determining molecular weights, in that an exact knowledge of the composition of the substance to be investigated, is necessary, that it may be used. We must know the simplest atomic composition, and also, approximately, the constitution of the compound. The number of double bonds, when they are present, must be known, and also whether it has a ring system. Also the valence of the nitrogen contained in it, must be known. Consequently, this method can be occasionally used to decide questions of constitution, in case the molecular weight is determined by another method.

The method of Traube gives the molecular weight of molecular complexes, for associated substances, therefore, the molecular weight of the simplest molecule, multiplied by the association factor. Consequently, it is often impossible to determine, from the result of a molecular weight determination alone, whether the substance has really a high molecular weight, or whether the molecular weight found is that of a molecular complex. It is, indeed, of less interest, for purely chemical investigations, to know whether the molecules of the substance investigated, are united homogeneously into more complex molecules, than to be able to find out the size of the simplest molecules which can exist. Whether a molecular weight found is really the simplest, or whether it corresponds to a somewhat more complex molecule, is shown by solution methods, at least with organic substances, from the course of the concentration curve. The question, as already observed, is not answered directly by the results of experiment, when the method of Traube is used, but through analogy, from the behavior of those substances which are most closely related chemically to the substance under investigation. But, since, from some knowledge of the substance to be investigated, it is almost always known whether more complex molecules are to be expected, the method of Traube can be used also to ascertain molecular weights for chemical purposes. The simplest molecular weight can, however, in many cases, certainly not be found by the method of Traube, as, for example, naphthalene. Important information pertaining to the derivation of the simplest molecular weights, is contained in the following pages.

DESCRIPTION OF THE METHOD OF TRAUBE

The method of Traube[1] was discovered and worked out in the years 1895 and 1896. It is purely empirical up to the present. It lacks theoretical foundation, a deficiency which it shares, moreover, with other important methods; e.g., the method for determining constitution from molecular refraction. The method of Traube leads to the molecular formula, from certain deviations obtained by determining, on the one hand, the molecular volume of the substance in question,

[1] J. Traube : Ztschr. anorg. Chem., 8, 323, 338 (1895); Ber. d. chem. Ges., 28, 410 (1895); Ann. Chem. (Liebig) 290, 410 (1896); Ber. d. chem. Ges., 28, 2722, 2728, 2924, 3292 (1895); 29, 1023 (1896).

directly, and on the other, by calculating it from definite known factors.

Experimental Determination of the Molecular Volume.— The space, measured in cubic centimeters, which a gram-molecule of a substance occupies, when undiluted with any foreign substance, is its molecular volume.[1] This includes the space present between the molecules of the mass. The molecular volume Vm, is obtained by dividing the molecular weight m, by the density d.

$$Vm = \frac{m}{d}$$

In order to be able to compare different substances in this regard, the density determinations have thus far been carried out under comparable conditions, and, indeed, at the boiling temperature of the substance under investigation. Traube chose a temperature of 15°, on practical grounds.

Calculation of the Molecular Volume.— Kopp had already recognized that the molecular volume is essentially an additive property of substances. Its value can be calculated, by adding the atomic volumes[2] of all the atoms which enter into the molecule. More thorough investigations have subsequently shown that the constitution, also, has an influence on the molecu-

---

[1] H. Kopp: Ann. Chem. Pharm., **96**, 153, 303 (1855).

[2] By atomic value is not to be understood the space occupied by the mass of the atom (volume of the atom itself), but the space necessary for the vibrating atom (space in which atom vibrates). According to Traube, the latter is about 3.5 times the former. Ber. d. chem. Ges., **29**, 2732 (1896).

lar volume. The presence of double bonds, the accumulation of halogen atoms on one carbon atom, etc., produce deviations in the value of the molecular volume, which cannot always be taken into account by an extension of the formula by which the calculation is made. Also, other irregularities manifest themselves.

Traube understood the reason for these deviations. He showed that the molecular volume cannot be calculated by simply adding the atomic volumes and some correction values, but that a numerical factor, which is always the same, must be added. The main point[1] of his work is this: *An increase in volume takes place when a molecule is formed from atoms; this is independent of the chemical nature of the substance, and can be only slightly modified by constitution. It amounts to 25.9 cc. for a gram-molecule at 15° C.*

This space of 25.9 cc. is designated as the covolume. The covolume represents the space which the molecule requires for its oscillations.

It is, indeed, very interesting that this covolume increases with the temperature, and, indeed, exactly according to Gay Lussac's law for the expansion of gases.

$$Cov_t = Cov_0 \ (1 + at) = 24.5 \ (1 + at).$$

Cov_t is the covolume at the temperature t; Cov_0 is the covolume at 0°; a is the coefficient of expansion of gases $= 0.00367$.

---

[1] J. Traube : Ann. Chem. (Liebig), **290**, 89 (1896).

Example:[1] The hydrocarbon $C_{13}H_{28}$ has a covolume 24.4 cc. at $0°$; 33.0 cm. at $100°$. The coefficient of expansion a_t is calculated from the equation $33.0 = 24.4 (1 + a_t t)$; therefore $a_t = 0.00353$. The deviation from the theoretical value is, as we see, very small.

The temperature correction is unimportant for the purpose of a molecular weight determination, and can be disregarded, if the density determination is carried out at the mean room temperature $(14°-17°)$. If greater differences in temperature occur, the correction can easily be made. The following short table contains the covolumes at several temperatures:

Temperature.	Covolumes.	Temperature.	Covolumes.	Temperature.	Covolumes.
0	24.5	15	25.9	22	26.5
5	24.9	16	25.9	24	26.6
10	25.4	17	26.0	26	26.8
12	25.6	18	26.1	28	27.0
14	25.8	20	26.3	30	27.2

It is clear, from what has been said, that the molecular volume can be obtained from the several atomic volumes, taking into account the covolume, and some correction values for any ring systems, double unions, etc., present.

The following table[2] contains the data necessary for the calculation.

[1] J. Traube : Ber. d. chem. Ges., **28**, 3297 (1895).

[2] J. Traube : Ber. d. chem. Ges., **28**, 2724, 2924 (1895) ; Ann. Chem. (Liebig), **290**, 43 (1896). In this paper the data observed are recorded : Ber. d. chem. Ges., **29**, 1024 (1896).

C	9.9	CN	13.2
H	3.1	N^{III}	1.5
O^{I}	2.3	N^{V} about 10.7	
O^{II}	5.5	N^{II} 8.5 to 10.7	
O^{V}	5.5	P^{III} about 17.	
O^{b}	0.4	P^{V} about 28.5	
S^{I}	15.5	Na	3.1
S^{II}	15.5	C_6 ring^{I} —8.1	
S^{II}	10 to 11.5	C_4S ring — 11.4	
F	5.5	N ring. small	
Cl	13.2	⊨	1.7
Br	17.7	≣	— 3.4
I	21.4		

In this table the symbols mean :

$O^{I}S^{I}$, hydroxyl oxygen, sulphydride sulphur. (Alcohols, etc.)

$O^{II}S^{II}$, the atom united with carbon by double union. (Ketones, Aldehydes, etc.)

O^{V}, an oxygen atom uniting two carbon atoms. (Ether.)

O^{b}, an oxygen atom in a carbonyl group, or which is united to a carbon atom, which is next to a carbon atom with a hydroxyl group attached to it. If there are more than one hydroxyl groups close together, one has O_1 oxygen atom.[2] (Acids, o-dioxybenzene, etc.)

$S°N°$, mean the atoms connected with oxygen (sulpho-, nitro-, etc. groups).

C_6 ring, denotes every ring of six carbon atoms.

C_4S, denotes the thiophene ring.

---

[1] Twice this value is to be subtracted in naphthalene, and three times the value in anthracene.

[2] Glycol has two oxygen atoms, one is to be introduced in the calculation as O^{I}, the other as O^{b}; a-oxy acids have O^{II}, O^{I}, Ob. So also β- and other oxy-acids.

N ring, means a ring system containing nitrogen.

⊨, denotes a double union.

⊫, denotes a triple union.

The values for the ring systems, and the double and triple unions, are, as negative values, to be subtracted from the sum of the atomic volumes.

If we designate by c the number of carbon atoms in the molecule of a compound, by h the number of hydrogen atoms, by O^I, O^{II}, O^V, O^h, the number of different oxygen atoms, by r the number of C_6 rings, etc., the molecular volume is :

$$Vm \quad 9.9\,c + 3.1\,h + 2.3\,O^I \cdots -8.1\,r \cdots + 25.9.$$

The molecular volume of a compound is calculated with the aid of this formula, which is easily completed from the foregoing table. Let some examples be given for this purpose. $\frac{m}{d}$ gives the value found experimentally.

Pentane, C_5H_{12}.		Benzene.			
C_5	49.5	C_6	59.4	C_6 ring =	8.1
H_{12}	— 37.2	H_6	18.6	3⊨	5.1
Cov.	25.9	Cov.	25.9		——
	——		——		13.2
	112.6		103.9		
$\frac{m}{d}$	113.7		—13.2		
			——		
			90.7		
		$\frac{m}{d}$	88.3		

Methylal, $C_3H_8O_2^v$.

$$C_3 = 29.7$$
$$H_8 \quad 24.8$$
$$O_2^v - 11.0$$
$$Cov. - 25.9$$

$$\overline{\quad 91.4 \quad}$$

$$\frac{m}{d} \cdot 88.4$$

Diallylglycerine ether. $C_9H_{16}O_2^vO^1 \models \models$.
Molecular volume for $0°$C.

$$C_9 \quad 89.1$$
$$H_{16} \quad 49.6$$
$$O_2^v \quad 11.0$$
$$O_1 \quad 5.3$$
$$Cov_0. - 24.5$$

$$\overline{\quad 176.5 \quad}$$

$$2\models \quad -3.4$$

$$\overline{\quad 173.1 \quad}$$

$$\frac{m}{d} \quad 172.6$$

DETERMINATION OF MOLECULAR WEIGHT

The molecular weight determination proper, depends upon this fact, that the molecular volume determined experimentally in the manner given, agrees with that calculated from the atomic volume, only when the correct molecular weight is assumed for the substance. Since, only then, does the following equation obtain:

$$\frac{m}{d} - \Sigma + 25.9 \quad \text{or} \quad \frac{m}{d} - \Sigma \quad 25.9.$$

By the sign Σ, is understood the algebraic sum of the atomic volumes and the correction values, with the exception of the covolumes. If the density is determined at a temperature (t), which differs considerably from $15°$, the value introduced is 24.5 $(1 + at)$, instead of 25.9.

If the molecular weight is estimated at one-half its real value, the equation will appear to be divided by two, and, accordingly, the numerical factor 12.9 appears instead of 25.9

If a molecular weight is assumed, which is double the real molecular weight of the substance, the equation will appear to be multiplied by 2, and the numerical factor upon the right will be 51.8. A few examples will serve to make the application of the method clear.

1. *What is the molecular weight of the chloride* $(CHCl_2)$, *whose density is* 1.6258?

The calculation gives the following values:

Formula assumed.	$\dfrac{m}{d}$	Σ	Difference.
$CHCl_2$	51.6	39.4	12.2
$(CHCl_2)_2$	103.2	78.8	24.4
$(CHCl_2)_3$	154.8	118.2	36.6

The examination of the column of differences, shows that the substance has the formula $C_2H_2Cl_4$, because the difference corresponding to this formula, 24.4, approaches closely to the normal covolume 25.9. The substance is tetrachlorethane.

2. *A bromide is obtained by introducing bromine into toluene, which, from analysis, has the formula* C_7H_7Br. *The density was* 1.401. *What is the molecular weight?*

Formula assumed.	$\frac{m}{d}$	Σ	Difference.
C_7H_7Br	122.1	95.5	26.6
$(C_7H_7Br)_2$	244.2	191.0	53.1
$(C_7H_7Br)_3$	366.3.	286.5	79.7

It follows from this calculation, that the simple formula C_7H_7Br belongs to the substance; for the covolume corresponding to this assumption, agrees closely with the required value 25.9, while the other differences are widely removed from it.

3. *What is the molecular weight of naphthalene* $(C_{10}H_8)_x$, *whose density is* 1.1517?

Formula assumed.	$\frac{m}{d}$	Σ	Difference.
$C_{10}H_x$	111.1	99.1	12.0
$(C_{10}H_8)_2$	222.2	198.2	24.0

It follows from this, that solid homogeneous naphthalene consists of double molecules.[1] On vaporization and solution, these molecules decompose to molecules of $C_{10}H_8$.

4. *Styrol, and two isomers formed from it, have the composition* $(C_8H_8)_x$; *the density of styrol is* $0.911 \left(\frac{15°}{4°}\right)$, *of the so-called distyrol,* $1.016 \left(\frac{15°}{4°}\right)$ *of the metastyrol,* $1.054 \left(\frac{13°}{4°}\right)$. *What are the molecular weights of the three hydrocarbons?*

---

[1] Naphthalene investigated by the freezing-point and boiling-point methods, showed not the least inclination to form double molecules. On the contrary, Küster had shown, both from the law of the division of substances between solvents, and also from the lowering of solubility, that solid naphthalene consists of double molecules. F. W. Küster: Ztschr. phys. Chem., 17, 366 (1895).

(a) Styrol.

Formula assumed.	$\frac{m}{d}$	Σ[1]	Difference.
C_nH_k	114.2	89.1	25.1
$(C_nH_k)_2$	228.4	178.2	50.2

Styrol has therefore the simple formula C_8H_8.

(b) Distyrol.

Formula assumed.	$\frac{m}{d}$	Σ[1]	Difference.
C_nH_k	102.3	89.1	13.2
$(C_nH_b)_2$	204.6	178.2	26.4

Distyrol has, therefore, the double formula, as has also been shown by the vapor-density determination. The high covolume suggests that when two styrol molecules unite, the union is effected through a double union of the side-chain in one of the molecules. The value of Σ, calculated on this assumption, would be 179.9, and the difference 24.7; this cannot be decided with certainty. As a matter of fact, distyrol takes up only one molecule of bromine, so that the conjecture appears to be justified.

(c) Metastyrol.

Metastyrol cannot be vaporized without decomposition, and is so slightly soluble in organic solvents that its molecular weight, thus far, could not be determined. The method of Traube, on the other hand, led to results.

Formula assumed.	$\frac{m}{d}$	Σ[1]	Difference.
C_nH_k	98.7	89.1	9.6
$(C_nH_k)_2$	197.4	178.2	19.2
$(C_nH_k)_3$	296.1	267.3	28.8
$(C_nH_k)_4$	394.8	356.4	38.4

[1] In calculating the sum Σ, for styrol and its isomers, a benzene ring and four double bonds are assumed to be present.

It is not very easy to decide in this case, since the value of the difference, which comes nearest to the normal covolume at $13°$ (25.7), differs from it by 3.1.

At any rate, three styrol molecules unite to form a metastyrol molecule; it is probable that two double unions in the side-chains are broken here, so that the formula of the compound is $C_{24}H_{24}$, $10\vDash$, $3C_6$ rings. The value of Σ, calculated for this, is 270.7; the difference thus being 25.4, while the normal covolume at $13°$, is 25.7 cc. A formula in which all the double unions of the side-chain are broken — indeed, a formula for which the great stability of the substance argues — is not possible, because in it a new ring system occurs.

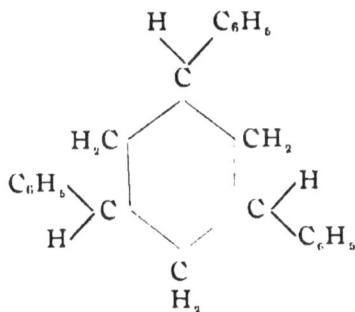

The value Σ, calculated for this, is 264.3; the difference would be 31.8; therefore, much too large. Deviations are found in a number of cases.

Indeed, larger covolumes sometimes appear even with the simplest possible formula, so that one could assume a partial decomposition of the simplest molecules. This anomaly, which has not yet been completely explained, is shown by substances into which more than one halogen atom has been introduced,[1] especially if the halogen atoms are united to one carbon atom; *c. g.*, chloroform (cov. $= 27.8$); chloral (cov. $= 29.0$); carbon tetrachloride (cov. $= 31.9$).

The same peculiarity exists with the highly substituted ammonias; *c. g.*, with triethylamine (cov. $= 31.6$); triisobutylamine (cov. $= 37.1$); and finally, with the ethereal salts of nitrous acid, and also — to a less extent—with those of nitric acid, if they contain a higher alcohol residue; *c. g.*, isoamylnitrite (cov. $= 27.7$); n octyl nitrite (cov. $= 30.0$).

This peculiarity must be taken into account in determining the molecular weight of the corresponding substances.

Example : Hexachlorethane has the density 2.011.

Formula assumed.	$\frac{m}{d}$	Σ	Difference.
C_2Cl_6	117.5	99.0	18.5
$(C_2Cl_6)_2$	235.0	198.0	37.0

Since substances with a large number of chlorine atoms have a high covolume, the conclusion must be drawn that hexachlorethane consists essentially of double molecules.

Another series of substances shows a remarkably small covolume, from which we must conclude that complex molecules exist. This includes substances containing hydroxyl, with low molecular weight, and also some other substances (as acetone) of very small molecular weight; *c. g.*, water (cov. 9.6); formic acid (cov. 15.5); acetic acid (cov. 18.7); valeric acid (cov. 21.7). The covolume also approaches the normal

[1] J. Traube : Ber. d. chem. Ges., **29**, 1027 (1896).

value with increasing molecular weight, as we see; *i. e.*, the association is less among the higher members of the series, as indeed, had been earlier made probable by vapor-density determinations. The following substances belong, also, with those of low covolumes: Methyl alcohol (cov. 15.6); ethyl alcohol (cov. 17.2); glycerine (cov. 14.9); acetone (cov. 19.0). The same rule, that the association decreases with increasing molecular weight, holds also for the alcohols; an increase in the number of hydroxyl groups in the molecule, increases the association factor.

The slight tendency of ethyl alcohol to associate is striking. We would conclude from the investigation of ethyl alcohol in benzene, by the freezing-point method (page 100), that there were molecules much more complex than double. It was, however, pointed out that that investigation was possibly not entirely to be depended upon, because, when the solution solidified, a mixture of alcohol with the benzene may have separated.

The *simplest* molecular weight of glycol, $C_2H_6O_2$, is found as follows, taking into account this influence of the hydroxyl groups.

Example: Glycol has the density $1.1279 \left(\dfrac{0^\circ}{4^\circ} \right)$.

Formula assumed.	$\dfrac{m}{d}$	Σ	Difference.
C_2H_6O	55	43	12
$(C_2H_6O_2)_2$	110	86	24

According to this, glycol in the liquid condition consists of double molecules. But since it contains more than one hydroxyl group, a covolume considerably smaller than usual, is to be employed in deriving the simplest molecular weight. Accordingly, the simplest molecular formula of glycol is $C_2H_6O_2$.

The *association factor*[1] (x) is derived from the co-volume (c), as follows: If c is found to be 25.9 at 15°, x = 1; if c is found to be 12.95, x = 2. If the value of c lies between 25.9 and 12.95, the association factor is found from the following equation :

$$x = 1 + \frac{25.9 - c}{12.95};$$

For example :

$$c = 19.1, \text{ then } x = 1.53.$$

Water has the largest association factor, 3.06; then comes glycol (1.88), glycerine and nitroethane (1.82), formic acid (1.80), acetic acid (1.56), methyl alcohol (1.79), ethyl alcohol (1.67). Small association is shown by benzene (1.18), and by toluene (1.08).

The Traube method of determining the molecular weight, can be very conveniently employed in many cases, indeed its experimental basis — a determination of the density — is easily carried out, at least for liquids. The density must often be ascertained by other determinations, perhaps refractometric, and then the molecular weight determination is limited to a short calculation. The Traube method gives, however, useful results only when an exact knowledge of the simplest atomic composition is obtained by analysis, and when the constitution of the compound is known exactly. It is absolutely necessary to know

[1] J. Traube: Ber. d. chem. Ges., **30**, 273 (1897).

the number and composition of the ring systems, which can be contained in the simplest formula, further, the valence of the nitrogen, and finally, at least approximately, the number of double and triple unions. The molecular weight of benzene, $c.\ g.$, could not be ascertained by the method of Traube, if the presence of the ring and the three double unions were not known. Such knowledge is not required to determine molecular weights by the other methods. They give the molecular weight of any substance, whatever, without requiring the least knowledge as to the composition, and have, accordingly, shown themselves to be especially useful in the many cases in which the beginning of a reaction is recognized most simply, and with the greatest certainty, by a molecular weight determination, without knowing anything as to the composition of the products.

MODIFICATION OF THE TRAUBE PROCEDURE FOR SOLUTIONS

It should be pointed out, in addition, that the Traube method can also be employed for ascertaining the molecular weight of dissolved substances. This is accomplished by deriving the molecular volume from the density of the solution. Differences of kind, depending upon the nature of the solvent, have become apparent, only indifferent solvents free from hydroxyl, showing no further anomalies; so that the molecular volume derived from the solution, just as that obtained from the homogeneous substance, can be introduced into the calculation. Solvents containing hydroxyl, especially water, lead, on the other hand, to entirely different results.

A. Indifferent Solvents. The molecular volume is derived from the density of the solution which is not too dilute, of the substance in chloroform, carbon bisulphide, benzene, or other indifferent solvent, in the following simple manner. From the volume of the solution which contains a gram-molecule of the dissolved substance, is subtracted the volume of the solvent contained in it. The difference is the molecular volume of the dissolved substance. If a solution of a gram-molecular weight of substance m, in 1 gram of solvent, has the density d, and the pure solvent the density δ, the molecular volume of the dissolved substance V_m, is expressed thus:

$$V_m \quad \frac{m + 1}{d} - \frac{1}{\delta}$$

Example: A solution of naphthalene in benzene, 18.77 per cent. had a density of 0.90312 at 19.1°; the density of the pure benzene is 0.8804.

The amount of benzene which would dissolve a gram-molecular weight of naphthalene (127.7 grams), giving a solution of the same concentration, is obtained from the equation,

$$18.77 : 81.23 = 127.7 : 1 = 552.6$$

$$V_m = \frac{127.7 + 552.6}{0.90312} - \frac{552.6}{0.8804} = 125.6.$$

The molecular volume of naphthalene, calculated by the method already given, on the basis of the simple molecule $C_{10}H_8$, is 125.3. On the other hand, a considerable difference exists between the molecular volume calculated, and that found, if the double formula $C_{20}H_{16}$ is made the basis of the calculation. Naphthalene dissolved in benzene consists, then, of simple molecules.

The molecular weight of a dissolved substance is also easily ascertained by the method of Traube, as is shown by this example. This modification of the method is of significance, especially for deriving the simplest molecular weight of solids, since it is difficult to determine accurately the density of solids. Inclusions of mother-liquor, air-bubbles, etc., introduce considerable sources of error into the direct determination of density, as is shown by the investigations of Retgers, published in the *Zeitschrift für physikalische Chemie*. The determination of the density of solids can also be avoided, when the substances have low melting-points, by determining the density above the melting-point, and including in the calculation a corresponding covolume. To determine the association factor of the solid, it is, of course, necessary to determine the density of the solid.

B. Aqueous Solutions.—The method of Traube applied to aqueous solutions, brings out very peculiar relations, which still need to be further studied.

It has been shown[1] that a contraction in volume takes place, when a gram-molecule of any substance, whatever, is dissolved in water. The molecular volume obtained experimentally, is about 13.5 too

[1] J. Traube : Ann. Chem. (Liebig), **290**, 88 (1896).

small. The equation which obtains for aqueous solutions is, therefore,

$$Vm' = \frac{m+1}{d} - \frac{1}{\delta} + 13.5.$$

It must, further, be taken into account that, strange to say, the double and triple unions are without effect on the molecular volume of a substance dissolved in water. Ring systems, however, have an effect. The sum, calculated in this way, using the atomic volumes already given, and the necessary correction values, is designated by Σ'. We have then:

$$Vm' = \Sigma' + 25.9$$

and further:

$$\frac{m+1}{d} - \frac{1}{\delta} = \Sigma' + 12.4.$$

Example:[1]

1 Phenol C_6H_6O, 3 $\mid=$. C_6 ring.	2 Allyl alcohol C_3H_6O', $\mid=$.
C_6 59.4	$C_3 = 29.7$
$H_6 -$ 18.6	$H_6 = 18.6$
$O' =$ 2.3	$O_1 =$ 2.3
Constants $=$ 12.4	Constants $= 12.4$
C_6 ring $= -8.1$	
	Vm' calc. $= 63.0$
Vm' calc. $=$ 84.6	Found $= 63.3$
Found $=$ 84.3	

It is clear that the molecular weight of substances can be calculated also on this basis, if the density of their aqueous solutions of definite concentration is known. The molecular weight is to be so chosen, that

---

[1] J. Traube : Ber. d. chem. Ges., **28**, 2726, (1895).

the calculated value of Vm' shall agree with that derived from the determination of the density, or that the value of Vm', obtained from the determination, is about 12.4 smaller than Σ'.

Alcohol.—A five per cent. aqueous solution of alcohol has the density $0.99029\left(\dfrac{15°}{4°}\right)$; a gram-molecule of alcohol (46 grams) requires 874 grams of water to form a five per cent. solution of alcohol. The density of water at 15°, is 0.99916.

$$Vm' = \frac{46 + 874}{0.99029} - \frac{874}{0.99916} = 54.3$$

Σ' is calculated as follows:

$$C_2 = 19.8$$
$$H_6 = 18.6$$
$$O = 2.3$$
$$\overline{\Sigma' = 40.7.}$$

The difference $54.3 - 40.7 = 13.6$, differs only a little from the required value. Therefore, the formula assumed, C_2H_6O, is correct.

From the density of a 10 per cent. solution of alcohol, $d = 0.98302\left(\dfrac{15°}{4°}\right)$, it follows that $1 = 41.4$, and $Vm' = 53.7$; the difference is 13.0. The deviation is here still smaller.

In consequence of the strong dissociation action of the water, no higher molecular weights were found for the substances containing hydroxyl, which were investigated in aqueous solution, as indeed is shown by ethyl alcohol. Furthermore, a large amount of halogen in the compound does not have any disturbing influence, so that with aqueous solutions results are always found which are easy to interpret.

On the other hand, electrolytic dissociation has an influence, since it increases the number of molecules, therefore decreasing the mean molecular weight. When there is *complete dissociation, according to Traube, the number 13.5 is to be subtracted from Σ', for the splitting up into two ions. Substances which undergo electrolytic dissociation must be investigated only in dilute solutions, since only such solutions can be regarded as completely dissociated; in more concentrated solutions the dissociation is driven back. Weak acids, which even in dilute solutions are only partially dissociated, are investigated in the form of their sodium salts, which are strongly dissociated in aqueous solution.

Example:[1] An aqueous solution of sodium m-amido-benzoate of 3.03 per cent. has the density $1.01266\left(\dfrac{15°}{4°}\right)$. $m = 159.11$, $l = 5092.04$, and $Vm' = 89.1$.

Σ' is calculated as follows:

$$
\begin{aligned}
C_7 &= 69.3 \\
H_6 &= 18.6 \\
N^{III} &= 1.5 \\
O_{II} &= 5.5 \\
O &= 0.4 \qquad\qquad C_6 \text{ ring } 8.1 \\
Na &= 3.1 \qquad \text{Dissociation constant } 13.5 \\
\cline{1-1}
&\ \ 98.4 \qquad\qquad\qquad\qquad\quad 21.6 \\
&- 21.6 \\[4pt]
\Sigma' &= 76.8
\end{aligned}
$$

---

[1] J. Traube : Ber. d. chem. Ges., **28**, 2730 (1895).

The difference 89.1 — 76.8   12.3, scarcely differs from the normal ∴ 12.4.   The formula assumed, $C_7H_6NO_2Na$, is therefore correct.

DETERMINATION OF THE DENSITY OF A LIQUID

To determine the density of a liquid, a small vessel with a narrow neck is filled up to a mark with water, weighed, then filled with the liquid to be investigated, and reweighed.   The vessel is also weighed empty.

By subtracting the weight of the empty vessel from each of the first two weights, the weights of equal volumes of water, and substance under investigation, are obtained.

Let m be the weight of the substance, w the weight of the water, q the density of the water at the temperature at which the measuring vessel was filled with it (to be taken from a table given in the appendix), λ the density of the air during the weighing, $\lambda = 0.0012$;

then we obtain the density of the substance by means of the following formula:

$$d = \frac{m}{w}\left(Q - \lambda\right) + \lambda.\,^{1}$$

The density, thus obtained, compares the substance at the temperature at which it was investigated, with water at 4° C.   The weighings are reduced to vacuum standard.

The temperature of the substance during the de-

---

[1] F. Kohlrausch: Leitfaden der praktischen Physik, Aufl. p. 61. (1896).   This formula suffices for density determination whose accuracy does not exceed one unit in the fourth decimal place.

termination, should then always be given with the
results of such a density determination.

It is also best to state that the density has been
referred to water at $4°$; perhaps thus, $\delta = 0.7983\left(\dfrac{16°}{4°}\right)$.
A modified formula is to be employed for density
determinations which should be accurate to the fifth
decimal place.

Pycnometers of different forms are used for weighing the liquids.

They have a capacity of from 1 to 10 cc. Determinations which are accurate to a few units in the fifth
decimal place, can be made with the larger kinds of
these pycnometers. Accuracy in the third decimal

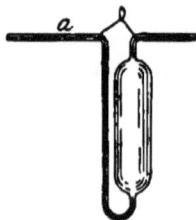

Fig. 44.
Ostwald's pycnometer.

place is sufficient for molecular weight determinations of homogeneous liquids; somewhat greater accuracy is required for solutions.

Small flasks, with a neck about 1 mm. wide, and
marked with a line drawn around them, are much
used as pycnometers. This pycnometer is not very
convenient to fill, because small air-bubbles easily
remain in it, which, even if they are very small, can
produce considerable error.

The Ostwald[1] form of the Sprengel pycnometer, shown in figure 44, is very convenient to use, and is to be recommended. The pycnometer consists of a small pipette, whose narrow tube is twice bent. It is provided, at A, with a mark which encircles the tube. The other tube is capillary. The liquid to be weighed, is drawn through the delivery-tube into the pycnometer proper, until this and the tube are filled up to the mark, by sucking on the capillary tube with an air-pump run by a current of water, or with the mouth, a calcium chloride tube having been placed between the mouth and pycnometer. The pycnometer is now completely freed from liquid particles clinging to the outside of the capillary etc., and is brought to the desired temperature. The liquid in the pycnometer is then so adjusted, that the capillary is completely filled, the other tube being filled up to the mark. This is accomplished by the following two modes of procedure: An excess of liquid is absorbed with a piece of cigarette paper placed at the end of the capillary, the liquid receding into the wider tube. If there is too little liquid in the pycnometer, a drop on a glass rod is brought to the end of the capillary tube filled with liquid. This is drawn in by suitably inclining the pycnometer, and then the final adjustment is made with paper.

The pycnometer, after it is regulated, is placed aside for some time, at a constant temperature, to see whether any changes in the temperature of the contents had taken place during the adjustment. To

---

[1] W. Ostwald : J. prakt. Chem., (N. F.), **16**, 396 (1877).

avoid such, the body of the pycnometer is never touched with the fingers, but it is always held by the tubes. The temperature is determined with a tested thermometer, since slight changes in temperature greatly affect the density.

The pycnometer is suspended on the balance during the weighing, by a small platinum hook placed above it. The weighing of the pycnometer, with the water, is carried out several times. The water used must be as pure as possible, and free from air. Water free from air is prepared by boiling it, and allowing it to cool in a space under diminished pressure. Such a determination gives the expression $\dfrac{Q-\lambda}{w}$, which is independent of the temperature. This expression is obtained with great accuracy by taking the mean of the individual determinations, which must differ only slightly from one another. These determinations are carried out as nearly as possible at the same temperature at which the pycnometer, filled with the substance, is later adjusted.

If a pycnometer is thus calibrated, then a density determination consists simply in filling the pycnometer with the substance to be investigated, and weighing it. The weight of the substance, multiplied by the above expression, and increased by λ, gives the density.

TENSION OF WATER-VAPOR.[1]

Temperature: t; Tension: w.

t	w	t	w	t	w	t	w	t	w	t	w
5.5	6.7	10.5	9.4	15.5	13.1	20.5	17.9	25.5	24.2	30.5	32.4
6	7.0	11	9.8	16	13.5	21	18.5	26	25.0	31	33.4
6.5	7.2	11.5	10.1	16.5	13.9	21.5	19.0	26.5	25.7	31.5	34.3
7	7.5	12	10.4	17	14.4	22	19.6	27	26.5	32	35.3
7.5	7.7	12.5	10.8	17.5	14.9	22.5	20.2	27.5	27.3	32.5	36.3
8	8.0	13	11.1	18	15.3	23	20.9	28	28.1	33	37.4
8.5	8.3	13.5	11.5	18.5	15.8	23.5	21.5	28.5	28.9	33.5	38.4
9	8.5	14	11.9	19	16.3	24	22.2	29	29.7	34	39.5
9.5	8.8	14.5	12.3	19.5	16.8	24.5	22.8	29.5	30.6	34.5	40.6
10	9.1	15	12.7	20	17.4	25	23.5	30	31.5	35	41.8

DENSITY OF AIR, OF WATER, AND OF MERCURY.

Temperature: t; Water of 4° as unit.

t	Air	Water	Mercury	t	Air	Water	Mercury.
1	0.0012883	0.99993	13.593	16	0.0012213	0.99899	13.556
2	836	97	591	17	171	0.99882	554
3	790	99	588	18	129	64	551
4	743	1.cc000	586	19	088	45	549
5	698	0.99999	583	20	046	25	546
6	0.0012652	0.99997	13.581	21	0.0012005	0.99804	13.544
7	607	93	578	22	0.co11965	0.99782	541
8	562	88	576	23	924	59	539
9	517	82	573	24	884	35	536
10	473	74	571	25	844	10	534
11	0.0012429	0.99964	13.568	26	0.0011804	0.99684	13.532
12	385	54	566	27	765	57	529
13	342	42	563	28	726	29	527
14	299	29	561	29	687	00	524
15	256	14	559	30	648	0.99570	522

[1] The numbers in this and in the following tables, are taken from the Physikalisch-chemischen Tabellen of Landolt and Börnstein.

VALUES FOR LOG $\dfrac{1}{1+at}$

Temperature : t ; $a = 0.00367$.

t	$\log \dfrac{1}{1+at}$	t	$\log \dfrac{1}{1+at}$	t	$\log \dfrac{1}{1+at}$
5.5	9.99132—10	15.5	9.97597—10	25.5	9.96115—10
6	99054	16	97522	26	96042
6.5	98976	16.5	97447	26.5	95969
7	98898	17	97372	27	95897
7.5	98821	17.5	97297	27.5	95824
8	98743	18	97222	28	95752
8.5	98666	18.5	97147	28.5	95680
9	98589	19	97073	29	95608
9.5	98512	19.5	96998	29.5	95536
10	98435	20	96924	30	95464
10.5	98358	20.5	96850	30.5	95392
11	98281	21	96776	31	95320
11.5	98205	21.5	96702	31.5	95249
12	98128	22	96628	32	95178
12.5	98052	22.5	96554	32.5	95106
13	97976	23	96481	33	95035
13.5	97900	23.5	96407	33.5	94964
14	97824	24	96334	34	94893
14.5	97748	24.5	96261	34.5	94823
15	97673	25	96188	35	94752

INDEX

ACCURACY in investigating very dilute solutions, increase
in 113
Acetic acid, vapor-density of, at different temperatures 50
Adjustment of the Beckmann thermometer 68
Alcohol in benzene, freezing-point lowering 130
Alcohol in glacial acetic acid, freezing-point lowering 139
Anomalies inherent in the freezing-point method 117
Anomalous results from vapor-density methods 49
Aqueous solutions, method of Traube applied to 221
Associating solvents 126, 192
Atmospheric pressure, effect of on boiling-point 184
Avogadro's hypothesis 1

BECKMANN boiling-point apparatus 145
Beckmann boiling-point apparatus, determination of molecu-
lar weights with 149
Beckmann's differential thermometer 66
Beckmann, freezing-point apparatus of 79
Beckmann freezing-point apparatus, carrying out a simple mo-
lecular determination with the 83
Benzoic acid anhydride in glacial acetic acid, freezing-point
lowering 140
Benzoic acid in benzene, according to freezing-point, and ac-
cording to boiling-point methods 194
Benzoic acid in benzene, freezing-point lowering 128
Benzoic acid in glacial acetic acid, freezing-point lowering 139
Benzoic acid in naphthalene, freezing-point lowering 128
Benzophenone in glacial acetic acid, freezing-point lowering .. 140
Boiling-point apparatus of Beckmann 145
Boiling-point apparatus of Jones 161
Boiling-point constant 142
Boiling-point constant, calculation of 144
Boiling jacket, modifications of the 170
Boiling-point method, determination of molecular weights by
the 141
Boiling-vessel, modifications of the 167

Boric acid in water, rise in boiling-point..................... 188
Borneol in benzene, rise in boiling-point..................... 193
Bott and Macnair, procedure of............................. 46
Burner, adjustable 174

CHOICE of freezing-point, or boiling-point method 196
Complex molecules 125, 191
Cooling vessel, influence of temperature of 116
Coste La, mode of procedure 45
Cresole, para, in naphthalene, freezing-point lowering.... 120, 136

DATA for calculating the density, by the Dumas method.... 39
Density of air, water, and mercury; tables................... 229
Densities of gases, determination of the 43
Densities of gases, determination of, under diminished press-
    ure... 44
Density of a liquid, determination of....................... 225
Determination of the boiling-point, with the Jones apparatus.. 164
Deville introduced the term "dissociation" 51
Dinitrophenol, o- and p-, in naphthalene, freezing-point lower-
    ing... 137
Dissociation electrolytic........................... 121, 189
Dissociation of vapors.................................... 49
Drops 17
Dumas and gas-displacement methods do not always give the
    same results 58
Dumas bulb 37
Dumas, method of 36
Dyson, procedure of..................................... 46

ELECTROLYTIC dissociation 121, 189
Electrolytic dissociation, measured by the freezing-point
    method... 123
Ethyl benzoate in benzene, rise in boiling-point.............. 187

FREEZING-POINT apparatus of Beckmann 79
Freezing-point apparatus of Beckmann, carrying out a simple
    molecular weight determination with the............... 83
Freezing-point method, anomalies inherent in 117
Freezing-point method, determination of molecular weights by
    the.. 73
Formanilid in benzene, freezing-point lowering 132
Formtoluide (o) in benzene, rise in boiling-point 195

Furnace, Lothar Meyer's 20

GAS-BURETTE 26
Gas-displacement method 6
Gas-displacement method, simple apparatus for the 8
Gas, filling the vaporizing vessel with an indifferent 27
Gay Lussac principle, methods based upon 34

HABERMANN 47
Heating box for boiling-point apparatus 148
Heat, source of 18
Heating the boiling-point apparatus 172
Hofmann modification of the Gay Lussac method 35
Hygroscopic solvents, procedure when they are employed 96
Hygroscopic solvents, stirrer used with 97

INDIFFERENT solvents 220
Inoculating rod 89
Iodine, vapor-density of 50

JONES, boiling-point apparatus 161
Jones, boiling-point apparatus, carrying out a determination
   with 164

LIQUID, determination of the density of a 225
Logarithm $\dfrac{1}{1 + at}$, values of 230
Lunge and Neuberg, the method of 45

MALFATTI and Schoop 47
Mechanical stirring device 93
Methylbenzaldehyde, m-, p-oxy-, freezing-point lowering 137
Methods based upon the Gay Lussac principle 34
Molecules complex 191
Molecular depression of the freezing-point 75
Molecules more complex 125
Molecular volume, calculation of 206
Molecular volume, experimental determination of 206
Molecular weight, derivation of from vapor-density 7
Molecular weights, determined by the freezing-point method. 73
Molecular weights, determined by the method of Traube 211
Molecular weights, determination of, by the boiling-point
   method 141

Molecular weights, determination of from the principle of lowering of solubility 197
Molecular weights, determination of with the Beckmann boiling-point apparatus 149
Molecular weights, determination of with the Beckmann freezing-point apparatus 83
Molecular weights of homogeneous solids or liquids 202
Molecular weight of solids, determination of 100

NON-ASSOCIATING solvents 126, 193
Naphthalene in glacial acetic acid, freezing-point lowering ... 118

OSMOTIC methods 62
Osmotic pressure 62
Oxybenzaldehydes, ortho, meta, and para, in naphthalene, freezing-point lowering 131, 135, 136
Ostwald's pycnometer 226

PEBAL'S diffusion experiment, showing the dissociation of the vapor of ammonium chloride 52
Perrot, gas-furnace for high temperatures 21
Phenanthrene in benzene, rise in boiling-point 188
Phenetol in glacial acetic acid, freezing-point lowering 119
Phenol in glacial acetic acid, freezing-point lowering. 139
Phosphorus pentachloride, dissociation of the vapor of 52
Pipette, weighing 90
Press for preparing tablets 100
Pressure, effect of atmospheric, on the boiling-point 184
Purification of solvents 178
Pycnometer, Ostwald 226

RAOULT developed the freezing-point method 74
Results from the boiling-point method 186
Results from the freezing-point method, critical examination of 117
Results of vapor-density methods 48

SCHALL, procedure of 47
Solids, determination of the molecular weights of 100
Solubility, determination of molecular weights, from the principle of lowering of 197
Solutions, increase in accuracy in investigating very dilute 113
Solutions, modification of the Traube method for 219
Solvents, associating 126

Solvents for boiling-point method 179
Solvents, indifferent 220
Solvents used in the freezing-point method 106
Solvents, use of the different, in the boiling-point method. .. 177
Solvents which cannot be used in certain cases, with the boil-
  ing-point method 181
Solvents which cannot be used in certain cases, with the freez-
  ing-point method 111
Stirring device, mechanical 93
Stirrer, effect of velocity of 115
Substance, introduction of into the boiling apparatus 175
Substances to be used in the vapor-jacket 18
Substance used in the vapor-density determination 22
Substances which form complex molecules 127

TABLET press 100
Tablet thrower 157
Temperature of the experiment, measuring the 32
Temperatures, the measurement of high 33
Tension of water-vapor, tables 229
Theory of a dissociating vapor 55
Thermometer, adjustment of 68
Thermometer, Beckman's differential 66
Thermometers, use of very large 115
Thermostat 102
Traube, description of method of 205
Traube method, as applied to aqueous solutions 221
Traube method as modified for solutions 219
Traube method, determination of molecular weight by 211

VAPOR-DENSITY, derivation of molecular weight from 1
Vapor-density determination, carrying out a simple 10
Vapor-density determination, theory of 2
Vaporizing-vessel, modifications of 24
Victor Meyer apparatus 9
Volume, calculation of the molecular 206
Volume, determination of the molecular 206
Volume of gas, determination of 24
WEIGHING glass 11

New Book on Physical-Chemical Methods.

⸙

The Freezing=Point, Boiling=Point,
and
Conductivity Methods.

⸙

BY

HARRY C. JONES,

Instructor in Physical Chemistry in Johns Hopkins University.

CLOTH   -   -   - $0.75.

"This book brings together in brief space the essentials of
theory and practice of these three methods, and is a valuable guide to students in laboratories of physical chemistry."—*Jas. Lewis Howe, in
Journal of the American Chemical
Society for April, 1898.*

⸙

Sent post-paid on receipt of price by

The Chemical Publishing Co.,
Easton, Penna.

STILLMAN.

Engineering Chemistry.— By THOMAS B. STILLMAN, Professor of Analytical Chemistry in the Stevens Institute of Technology. A Manual of Quantitative Chemical Analysis for the use of Students, Chemists, and Engineers - - - - $4.50 8vo. Cloth. 154 Illustrations.

VENABLE.

The Development of the Periodic Law.— By F. P. VENABLE, Professor of Chemistry in the University of North Carolina. - - - - - - - - - - - - $2.50 12mo. Cloth. Well Illustrated.

VENABLE AND HOWE.

Inorganic Chemistry According to the Periodic Law.— By F. P. VENABLE, University of North Carolina, and JAS. LEWIS HOWE, Washington and Lee University - - - $1.50 Cloth. 35 Illustrations.

WILEY.

Principles and Practice of Agricultural Chemical Analysis.— By HARVEY W. WILEY, Chemist of the U. S. Department of Agriculture. Complete Work, Bound in Cloth, $9.50; Bound in Half Morocco - - - - - - - - $12.50

Vol I.—Soils - - - - - - - - $3.75
607 pp. 93 Illustrations, including 31 Plates.

Vol II.—Fertilizers - - - - - - - $2.00
332 pp. 17 Illustrations.

Vol III. Agricultural Products - - - $3.75
665 pp. 125 Illustrations including Plates.

Methods for the Analysis of Ores, Pig Iron, and Steel, in use at the Laboratories of Iron and Steel Works in the Region about Pittsburg, Pa., together with an Appendix containing various Special Methods of Analysis of Ores and Furnace Products. Contributed by the Chemists in charge, and Edited by a Committee of the Chemical Section, Engineers' Society of Western Pennsylvania - - - - **Paper, $ .75; cloth, $1.00**

The Chemical Publishing Company,

EASTON, PENNA.